TECHNOLOGY STARTUP NARRATIVES

Murali Nair

Disclaimer:

The views expressed in the book are those of the author and do not necessarily represent the views of the National Science Foundation of the United States

my beloved gudiya, my inspiration

CONTENTS

FOREWORD 11

INTRODUCTION 13

NARRATIVES 23

1	Robots Picking Lettuce	25
2	Cleaning Water with Light	39
3	I Can Compute Emotion	53
4	An Even Smaller Memory Chip	69
5	Oceans Provide Sustenance	83
6	No Need for Too Much Lubricant	97
7	A Sense of Touch	111
8	What I Wear Harvests Energy	127
9	Let Us Not Allow Milk to Spoil	143
10	Helping Wheelchairs Navigate	159
11	You Can Print Batteries?	173
12	Down with Internet Hackers	187
13	A Spectrometer in Your Hand	201

14	Chips Unable to Keep Time	215
15	Sensual Touchscreens	231
16	No Need for Spiders to Make Silk	245
17	Plastics Transmit Radio Waves	259
18	Displays You Can Bend and Twist	273
19	Measuring What You Cannot See	289
20	Plastic Baby Bottles Can Be Safe	301
21	Fabrics That Can Remember	315
22	Can You Believe I am Walking?	329

AFTERWORD	**343**
ACKNOWLEDGEMENT	**345**
REFERENCES	**347**
ACRONYMS	**349**

FOREWORD

Descartes[1] said that systematic doubt is the first principle of learning. In the seventy-five years since the classic characterization by Schumpeter[2], of the 'creative destruction' process of industrial technological change, the role of technology in economic growth has grown relentlessly. Two major research and development (R&D) policy response mechanisms are tax policy and direct funding. The policy models and analyses presented are based largely on U.S. economic experience, but the resulting prescriptions are relevant for all existing and emerging technology-based economies. The industry-centric Schumpeterian model must be expanded to one in which competition among governments is as important as it is within the private sector.

The thesis in the rise and decline of nations[3] was that decline comes because after many years of success, that a nation's political and economic arteries get so clogged with special-interest groups that its life-giving circulation of ideas and elites is impaired. The political and governmental entrenchment and decline leads to elites that are calcified and verging on permanence. Society runs out of creativity which limits innovation and complicates any prospect of renewal. An economy where manufacturing is twelve percent, the financial sector is twenty percent, and the financing of housing is twenty-five percent lets itself luxuriate in finance at the expense of harvesting, manufacturing, or transporting things. This lets a financial elite elevate, expand, and entrench itself as a country's Gross National Product (GNP) (and profits) dominating center. Debt and the emergence of the U.S. financial sector provides a revealing connection.

The only necessity for evil to prevail is for good men to do nothing; will good men and women do little, or will they respond to the emergency that is now at hand[4]? The convergence of technology-based competitive capabilities among the world's economies has drastically altered the required economic growth strategies in industrialized nations. Based on a variety of corporate and government investment trend data and comparisons among national growth strategies, Tassey[5] examines how this convergence has created an imperative for new growth models and strategies and analyzes the major policy mechanisms for stimulating R&D investment and improving R&D efficiency over technology life cycles, detailing the needed changes.

A previous volume[6] detailed a possible methodology to successfully transform public investments in basic and applied research to economic value and societal impact through federal funding of technology startups that facilitate innovative business models through coordination and partnership with the private sector. For over three decades, programs across several federal agencies have provided funds to tens of thousands of startups and helped them bring new products and services to market.

These programs nurture innovation for economic and societal benefit by investing in industry-transforming science and engineering research and development. By promoting collaboration between industry and academe it has helped create successful technology startups and increased long-term, leading-edge United States' innovation capacity.

Like the Harvard Business School case studies that focus for the most part on large corporations, this volume seeks to do the same for high-technology startups that were initially funded by the National Science Foundation's Small Business Innovation Research Program. The original intent to build a repository of hundred such case studies remains. In this volume, twenty-two narratives of high-tech startups, all recipients of grants from this program, cover the entire gamut of technologies arising out of basic research investments, from life sciences to physical sciences, from synthetic biology and medical devices to nanotechnology and sophisticated instruments. These feature the different and varying aspects of current technology development trends such as for instance the interplay of materials, biology, and information technology innovation.

Murali Nair
December 31, 2017
Virginia, USA
muralinair@yahoo.com

INTRODUCTION

A nation that capitalizes on advances in science and engineering and provides global leadership in research and education will increase its productivity and thus prosper. Transformative research leads to the emergence of new fields and extraordinary shifts in existing fields and, by its nature, has significant impact on the frontiers of science and engineering. Research and innovation are the spark for creating whole new industries, such as the Internet, search engines, tissue engineering, the global positioning system, cellular biology, 3-D printing, and touchscreen technology. The United States' National Science Foundation (NSF) supports the basic research that enables advances ranging from technology-based innovations that spur economic prosperity to understanding, mitigating, and adapting to climate change to developing sustainable approaches to the utilization of energy and natural resources. By forging links between fundamental research and society's needs, this federal agency helps articulate important new areas of science and engineering, improves the quality of life, creates a scientifically literate populace, and empowers future generations. Additionally, NSF has supported the research efforts of over two-hundred and forty Nobel Laureates in their early research careers.

Revolution is not achievement and the new dawn; it results from senile decay, from the bankruptcy of ideas and institutions, and from the failure of self-renewal. Politics is still based on the age-old assumption that whatever government does is grounded in human society and therefore "forever"; as a result, there is no political mechanism in government to slough off the old, the outworn, and the no-longer productive. Politicians and the bureaucracies over which they preside love power, and power is never easily surrendered. It was Lincoln who said that nearly all men can stand adversity, but if you want to test a man's character, give him power.

Some with slim track records are entrusted with too much too soon; talent is indeed essential but seasoning and maturity are not to be underestimated. Leadership is encouraging people to exceed their own expectations; in inspiring people to be great; and in getting them to do it all together, in harmony. Hierarchies are just too slow, too rigid, too self-obsessed, too mired in a sort of perpetual bad mood. And most of the time, their leaders are in over their heads. Management is a methodical process. Its purpose is to produce the desired results on time and on budget. It complements and supports but cannot do without leadership, in which character and vision combine to empower someone to venture into uncertainty. Leaders must suspend the disbelief of their constituents and move ahead even with incomplete information. Managing, vigorously driving execution, is a rare skill, but rarer still is the ability to lead, inspire, and motivate people.

Deciding which opportunities should make the transition to incubation once they have been developed and articulated is a key function of discovery leadership. While the primary focus of NSF-supported research is the generation of new knowledge, these programs, where appropriate, should consider stakeholder input to optimize the utility of research to address societal challenges. NSF is committed to creating connections between research produced through its investments and the needs of society. This requires close interaction with stakeholders and a clear recognition of the Foundation's role in the nation's innovation enterprise. Institutions that are critical to innovation must be supported by expanding funding not just for university research, but for the kinds of mechanisms and institutions that help foster commercialization of research through viable business models. Many important innovations present serious challenges in mass production or mass distribution. Others require considerable infrastructural investment or coordinated investments by many different entities. It is critical to use the federal government's coordinating abilities to overcome information mismatch problems to bring together innovators and markets.

The national and global economic landscape has undergone vast alterations over the past few years. The nation must maintain a robust science, technology, engineering, and mathematics (STEM) workforce. Federal programs must consider the international opportunities and challenges presented by the dynamic global research and learning environment. Another major factor influencing the conduct of twenty-first century research is the emergence of massive amounts of data and the changing capacity of scientists and engineers to maintain and analyze it. We must catalyze the education, for example, of technologically literate entrepreneurs, and making the innovation process, from fundamental research breakthroughs to products and services that will benefit society, more robust and effective. It is necessary to partner with industry to develop overarching solutions to pernicious worldwide blights and global challenges, and to reach out to the communities that play complementary roles in the innovation process.

The Small Business Innovation Research (SBIR) program is a U.S. federal program, coordinated by the Small Business Administration (SBA), intended to help certain small businesses conduct research and development (R&D). Funding takes the form of contracts or grants. The recipient projects must have the potential for commercialization and must meet specific U.S. government R&D needs. This program was created to support scientific excellence and technological innovation through the investment of federal research funds in critical American priorities to build a strong national economy. It is to provide funding for some of the best early-stage innovation ideas, ideas that, however promising, are still too high-risk for private investors, including venture capital firms. The term "small business" is defined as a for-profit business with fewer than five hundred employees, owned by individuals who are citizens of, or permanent resident aliens in, the United States of America.

Funds are obtained by allocating a certain percentage of the total R&D budgets of the eleven federal agencies with extramural research budgets that exceed $100 million per year. Approximately $2.5 billion is awarded through this program each year. The mission of the SBIR program is to support scientific excellence and technological innovation through the investment of federal research funds in critical American priorities to build a strong national economy. The program's goals are four-fold:

1. Stimulate technological innovation.
2. Meet federal research and development needs.
3. Foster and encourage participation in innovation and entrepreneurship by women and socially or economically disadvantaged persons.
4. Increase private-sector commercialization of innovations derived from federal research and development funding.

NSF SBIR

The NSF SBIR program is administered through the Division of Industrial Innovation and Partnerships (IIP) in the Directorate of Engineering. One key driver of this program is to advance its topics for submission of proposals, to beyond the state-of-the-art. The NSF SBIR program should be in the vanguard of future trends and developments even as a part of the federal government where hierarchies are impenetrable and micromanagement ineffective. With their mania for control, bureaucracies must learn to cope with a new world powered by innovation and creativity. Do we seed whole new industries in America, or do we simply continue to emulate "spray and pray" venture capitalists (VC) and their attendant scatter-shot approach?

The pilot SBIR program was first implemented at NSF in 1978. The NSF SBIR program size in 2017 is $185M per year with an annual volume of Phase I proposals submitted to the program averaging approximately 2,500. At any given time, there are about 350 Phase II proposals in the pipeline. Phase I is feasibility research conducted over a period of between 6-12 months, with a maximum award size of $225K. Phase II is research towards development of a prototype of a product, process, or service, followed by additional investment and/or partnering with the private sector for market entry. The maximum Phase II award size is $750K for a period of two years or less, with additional supplements possibly worth another $750K. All Phase I and Phase II proposals are peer-reviewed by panels of experts from both academia and industry from around the country. All panels convene at NSF. On average, the amount of private capital raised by NSF SBIR grantees every year is between $30M and $50M. The following Table provides examples of companies that have been funded by this program since its inception in 1980.

Company	Founded	Description	Market Cap
Blue River	2012	Robot to thin, weed, and pick lettuce using computer vision and machine learning. Acquired by John Deere in 2016	$305M
ecoATM	2008	Automates inspection, pricing, payment, and take-back of used consumer electronics, 270 employees, 1100+ kiosks in 42 states; acquired by Outerwall in 2013	$350M
Intralase	1997	California-based developer of femto-second laser technology, acquired in 2007 by Advanced Medical Optics, a maker of medical devised and laser vision-correction systems	$808M
Qualcomm	1985	Andrew Viterbi and six colleagues formed Qualcomm in 1987. NSF SBIR provided $265K for single chip implementation of Viterbi decoder; led to high-speed data transmission via wireless and satellite networks; holds more than 10,100 U.S. patents, licensed to more than 165 companies	$78B
Symantec	1981	Gary Hendrix founded the company: NSF SBIR awarded $30,000 to develop a framework for managing dissimilar data; now makes security, storage, and backup software: most-used certification authority	$12B

The program funds small businesses developing high-risk, high-payback innovations that demonstrate strategic partnerships with research collaborators, customers, and equity investors, and demonstrate the potential for marker acceptance. This program does not fund evolutionary optimization of existing products and processes; or modifications to broaden the scope of an existing product, process, or application; or analytical or market studies of technologies. The NSF SBIR program also applies a societal impact review criterion, an example of which is the development of enabling technologies such as research tools and instruments. This program invests in risky technology in a relatively conservative way. It is meant to be open and broad in terms of the technologies supported.

Areas include but are not limited to:

- Biological, biomedical, environmental, and chemical technologies

- Educational technologies and applications

- Electronics, information, and communication technologies

- Nanotechnology, advanced materials, and manufacturing

The NSF SBIR program helps to de-risk innovative, risky early-stage innovations for later private sector investments, and helps startups bridge the so-called first valley of death in the discovery-development-commercialization pathway to market entry. There is also a second valley of death – the woes associated with scaling, supply-chains, and manufacturing in volume. NSF program officers are engaged in business and technological brokering by linking university researchers to entrepreneurs interested in starting a new firm, connecting startup firms with venture capital, finding a larger company to commercialize the technology, or assisting in procuring a government contract to support the commercialization process.

Most thinkers on strategic management and organizational change have focused more on the private sector, leaving the public sector to simply focus on 'creating the conditions' for innovation to happen in the 'revolutionary' private sector. We need to transform the public sector from within so that it is more strategic, meritocratic, and dynamic. While government spending on basic education and health should not necessarily expect a direct return beyond the taxes and supply of skilled and healthy staff, the government's high-risk investments should be thought of differently, and allowed to reap a direct return precisely because the failure rate is so high.

NSF SBIR Review Criteria

Two overall review criteria are used to assess proposals - intellectual merit and broader impact.

Intellectual Merit

Is the proposed plan a sound approach for establishing technical and commercial feasibility? To what extent does the proposal suggest and explore unique or ingenious concepts or applications? How well qualified is the team to conduct the proposed activity? Is there sufficient access to resources? Does the proposal reflect the state-of-the-art in the major research activities proposed? Are advancements in the state-of-the-art likely?

Broader Impact

What may be the commercial and societal benefits of the proposed activity? Does the proposal lead to enabling technologies for further discoveries? Does the outcome of the proposed activity lead to a marketable product or process? Does the proposal evaluate the competitive advantage of this technology versus alternate technologies that can meet the same market needs? How well is the proposed activity positioned to attract funding from private sources once the SBIR project ends? Can the product or process advance NSF's goals in research and education?

Additional criteria that are used to assess each proposal involve the following: the market opportunity; the team; the technology and competition; and the revenue and finance plan.

Market opportunity

Does the proposal succinctly describe customer need, and how this product or service aims to fulfill that need? Is the business model for this innovation clearly defined – does it involve a service, product, or license? What is the market entry strategy? Where is the technology placed in the development cycle? Are the critical milestones required to get to market clearly defined? Is there a strategic reason for the nation to invest in this innovation?

Team

What is the vision of the founders, and their background/ experience? Does the company have access to useful networks of investors and/or strategic partners? How well is the team poised to take this innovation to market? Have they taken comparable products to market previously? Does the company have additional outside advisors, mentors, partners, and stakeholders? Is the corporate structure consistent with the company's vision?

Technology and Competition

Did the company describe the features of their technology? Does it provide a compelling value proposition to the customer? Have they validated the proposed value proposition? What will a customer pay for the innovation or service enabled by the innovation, and is this validated? Does the company demonstrate a realistic understanding of the cost to take this innovation to market? Who are the main competitors in the market space? Does the company clearly describe how they are going to compete: price, performance or other? Has the company adequately addressed the intellectual property (IP) landscape and its position in this domain? Any evidence that a patent search has been undertaken? Is there an IP strategy?

Revenue and Finance Plan

What is the company's revenue history and revenue sources, if any? Does the company demonstrate adequate knowledge of the level of financial resources required to get the innovation to market? Do they have a phased plan to bring these funds to the table? Is there evidence of commitment for the funding beyond SBIR funding? Are the revenue streams for this innovation described? When will those revenues be received? When does cashflow break-even occur? Is the pro-forma reasonable given the state of the innovation? Has the company adequately described and tried to validate the assumptions behind the models?

Of the twenty-two narratives that follow, in twelve of them, the NSF SBIR program provided additional commercialization assistance to the startup. In eleven cases, multiple interviews were conducted with the founding team to construct the narrative. The Table below provides more granularity by technology sectors, locations, and business/growth models of these narratives.

Startup	Technology Sector	State	Business/Growth Model
Blue River	Computer Vision	California	VC
Dot Metrics	Optoelectronics	North Carolina	M&A
Affectiva	Software	Massachusetts	VC
Zeno	Semiconductors	California	Strategic Partnering
Stellar	Biotech	California	IPO
Fusion	Manufacturing	Michigan	Strategic Partnering
NextInput	Displays	Georgia	VC
Perpetua	Wearables	Oregon	Strategic Partnering
Promethean	Materials	Massachusetts	Bootstrap
Love Park	Robotics	Pennsylvania	Bootstrap/ Strategic Partnering
Imprint	Energy Storage	California	VC/Strategic Partnering
BitSight	Cybersecurity	Massachusetts	VC
Active Spectrum	Instrumentation	California	Bootstrap/M&A
Blendics	Semiconductors	Missouri	Strategic Partnering/Bootstrap
Tanvas	Displays	Illinois	VC/ Strategic Partnering
Bolt Threads	Synthetic Biology	California	VC
NetBlazr	Wireless	Massachusetts	Diversification/Bootstrap
Orthogonal	Optoelectronics	New York	Strategic Partnering
Anasys	Instrumentation	California	Bootstrap/ Strategic Partnering
Plastipure	Life Sciences	Texas	Bootstrap/ Strategic Partnering
MedShape	Medical Devices	Georgia	VC
Ekso Bionics	Robotics	California	IPO

The Table indicates that a significant percentage of awards are garnered by California and Massachusetts, with a sprinkling across nine other states. Sometimes, a startup is forced to relocate because of a lack of a thriving ecosystem and the required technology infrastructure. In one case presented in this volume, a startup, although initially located in Georgia, found that as it matured, the company was forced to relocate to California for reasons of better access to venture capital funding sources and the availability of appropriate human capital. The last column in this Table lists the various business/growth models adopted by this set of startup companies – use of venture capital (VC); mergers & acquisitions (M&A); initial public offering (IPO); strategic partnering/joint development agreements (JDA); bootstrap growth via sales, contracts, and licensing; diversifying from an existing line of business; or sometimes a mix of these. There are also instances in these narratives of pursuing a different sub-market than originally intended or the technology pivots to address the needs of the originally intended market segment. It is important to note that the NSF SBIR program itself is entirely technology- and location-agnostic.

Finally, narrative 12 provides an example of the garage-to-market innovation model and narrative 18 similarly provides an example of the lab-to-manufacture innovation model. At the end of each startup narrative, a set of case study-type questions is provided for further discussion.

NARRATIVES

1   *Robots Picking Lettuce*

Technology Sector: Computer Vision
Startup: Blue River Technology, Inc.
Website: www.bluerivert.com
Location: Sunnyvale, California

Federal Funding Timeframe: January 2012 – February 2018
Funding Amount: $1.15M

This narrative is a story of: The use of private equity funds beyond federal funding; Mergers & Acquisitions (M&A): quick exit (less than five years) via acquisition by John Deere, Inc.; significant societal impact

Startup Formation

Lee Redden, one of two founders of Blue River Technology, Inc. ("Blue River"), grew up in Nebraska, where he attended the University of Nebraska at Lincoln to initially pursue a degree in mechanical engineering. One hour away from Lincoln, his uncle farmed 6,000 acres of corn and soybeans. Lee remembers visiting this farm on family vacations – he loved the machines, the variegated farm activities, and rhythms. He thought to himself - so many toys to play with, all mechanized. He was impressed that it took only six people to cultivate and harvest the produce from 6,000 acres, so amazingly efficient! In junior year Lee switched to a more robotics-oriented program with electrical engineering and computer science now his focus. After completing his undergraduate degree, now deeply interested in robotics, he found positions at National Aeronautics and Space Administration (NASA) Johnson Space Center where he worked on a simulator for astronauts to acclimatize them to zero-, lunar-, and Martian-gravity conditions and then at Johns Hopkins University Applied Physics Lab. A mentor at NASA often discussed with him robotics research at Carnegie-Mellon University (CMU), Massachusetts Institute of Technology (MIT) and Stanford University ("Stanford").

Lee applied and obtained a National Science Foundation (NSF) graduate research fellowship at Stanford. Lee's goal was to obtain a doctorate in robotics but after finishing his master's degree, he took a leave of absence from his doctoral work to devote all his time to Blue River. At Stanford, he shared an apartment with three other engineers, all intent on starting companies. Their days were filled with many conversations about it, innumerable technologies they were interested in and wanted to pursue, multiple exciting possibilities – these discussions molded Lee's attitude and perspective enough to change his mind about pursuing a doctorate. He gravitated to precision agriculture and took a class on 'lean startups' where he met Jorge, the other founder of Blue River. In the 10-week class they joined a team building an autonomous lawn mower –throughout the course they visited agricultural farms and were especially intrigued by the weeding of large carrot fields, and then lettuce thinning. Jorge and Lee built out the computer vision aspect of the team project. The duo started Blue River in 2011 with the initial focus on building general-purpose computer vision applications. Jorge has deep experience with agricultural technology companies, at Trimble Navigation, where he led the precision agriculture group, and is an expert in finance and M&A while Lee had developed expertise in computer vision and robotics – agriculture thus the common element between the two founders.

When asked about role models and the company vision Lee said – "There were many mentors from the U.S. space program who had developed all kinds of airplanes and rockets. NASA was inspirational – I learnt so much there. My undergraduate advisor was also a role model - he founded a few companies based on his research. I liked that model, a place for research to go. If there is a vision that guides me, it is to bring computer vision and machine-learning to agriculture, machines operating on farms will have information on individual plants, personalized medicine for plants, plant-by-plant care."

Technology

Lee developed the original technology platform from the different methodologies and techniques[7] he had learned at Stanford and started to build out specific applications to precision agriculture with this platform as the base. This was not a part of his laboratory work at Stanford while he pursued his incomplete doctoral research – at any rate the computer vision lab at Stanford is not generally concerned with agricultural applications. This meant that Blue River did not have to license technology from Stanford. All intellectual property created thus far was generated by work at the company and belongs to Blue River. The firm has filed four provisional patents and is starting to convert these to seven U.S. patents - possibly two of these with an option to convert to international patents. The innovation tailors' machine-learning components to real-time tasks pertinent to precision agricultural applications to elicit required plant features. The many algorithms used in machine-learning traditionally use web-based data sets – the challenge is to adapt these to agricultural applications and modify them for real-time execution. These algorithms have also to be extremely robust.

Computer vision routines serve well as research publications but are not robust enough for time-varying field conditions. This technology must capture many different sets of parameters – for example, plant visual characteristics such as its size, color, and location, to properly identify lettuce in the field. Currently there is a company in Salinas, and another in San Diego that are developing automated lettuce thinners that the founders maintain do not work all that well. The secret is to build machines that perform exemplarily well as defined by farmers. The key technical milestones for the next two years are to focus and expand on the lettuce-thinning application. Blue River's current machine works as expected especially on the algorithmic side. Field tests are going well. Machine design and final assembly are done in-house, while different machine components such as the wiring harness, printed circuit boards, and other standard parts are outsourced to shops in California. The firm partners with farmers for use of their fields for testing. The value proposition for farmers is an automated process to remove plants that are not required, and simultaneously not destroy lettuce plants that the farmer wants to harvest without having to hire large labor crews that are either unavailable or are expensive.

Specific team members are assigned to the third and fourth generations of the machine and others like Lee are involved in the design of the fifth-generation machine. To diversify into markets beyond lettuce, two research projects will be pursued in 2015-16. The first will address phenotyping, the observable physical or biochemical characteristics of an organism, as determined by both genetic makeup and environmental influences, and the second 'weeding'. The company is starting to tackle the weeding issue – interviewing farmers about their weeding requirements, collecting different data sets, and researching weeds particular to lettuce, for different plants have different sets of weeds. For general purpose weeding, the idea is to find a few crops that have a common set of weeds. Phenotyping would entail individual plant measurements for breeder trials to generate better analytics for a variety of vegetable crops – tomatoes, cucumbers, melons, flowers, canola, sweet corn, and field corn, the last an exciting big market. Lee is responsible for the technology roadmap, to determine realistic areas for the company to pursue. He realizes that phenotyping would require extensive knowledge of plant types, and an additional team headed by a renowned agronomist.

Customer Discovery and Business Model

How did you find customers?

"Jorge makes the initial contact. The two of us then together talk to the farmer in his field – we start by showing them images of their crop. We discuss the unique needs they have – they guide us, tell us what they really want. Can we do it, they ask. Later we show farmers field images to demonstrate that we can deliver technology solutions that address their needs. Farming culture is a show-me culture; fancy presentations don't work. Our first customer was initially skeptical but later after we demonstrated the machines in the field found the technology credible. Our first customer produces approximately 10% of the lettuce grown in the U.S. Fifty per cent of the lettuce grown in America is produced in California's Salinas Valley or Yuma in Arizona by seven growers, we maintain good relationships with them, and have signed up six of the seven. They all want our technology to work and work robustly. We then make a formal proposal – how they can increase their lettuce yield with fewer inputs. For example, a 25-acre field requires a labor crew of 25-30 people for a day for thinning not harvesting; harvesting requires more labor. We present the value proposition - all this is mechanized with minimum manual labor for thinning; no outside crew to hire. Our field tests can convince the farmer."

Does this not eliminate jobs?

"It doesn't, there is a shortage of labor crews with the problems along the border. If the economy is good, people prefer to do construction, rather than this unglamorous, backbreaking work. This is for them not a career – they usually do lettuce thinning for a year then find jobs in other industries. It is why there's a big labor shortage. We are not in the business of eliminating jobs. We want to make it more convenient to not depend on too much hard labor."

After convincing the first customer, the second to the sixth customer has been a much easier sell. Later sales involve persuading growers using the pros and cons of machine versus manual, and Blue River's machine versus another competing automated service. There is a local company in Salinas building machines for lettuce thinning. Customer feedback suggests that the performance of Blue River's machine far surpasses it.

Will this put them out of business?

"No, they make lots of different equipment, so it won't have a large effect on them. Automated machines currently make up a small percentage of total thinning, so there is a lot of room for many companies to grow."

Why pick lettuce? What is your business model?

"It's a big market – it is harvested every single day for it grows throughout the year, a wonderful crop! The business model is a two-fold approach; the first is an increase in yield, around 10% for lettuce using our machine versus employing manual labor. We also help optimize the lettuce left in the field after thinning by packing them closer together; from the information gathered and images seen by our machine, we can decrease the standard deviation of plant field distribution the normal being eight-nine inches. The second is labor – farmers find it hard to find timely labor, moreover, a different crew is required for different operations – thinning, harvesting – we reduce that need. We charge a premium per acre over manual crews that include scheduling and logistics of a truck, trailer, tractor, and our machine, one machine for a 25-acre farm. We do all the work end-to-end, and the farmer doesn't need to do anything. It's a bargain for the farmer who gets a field with higher yields. The same farm uses crop rotation and grows broccoli, cauliflowers, carrots, and tomatoes. To do all five is a big deal with many different operations. Harvesting provides a vast area for expansion, but it is very difficult to do with a machine."

Traditional methods of thinning using manual labor remain the competition. Farmers are comfortable with that and normally fear the unknown, and manual labor when available is very reliable. Blue River's machine performs much better, at least two times better. The machine is good at repetition, but when it does wrong, it keeps doing wrong. A second camera, a back camera, was added to solve this problem by looking back at what was sprayed. This provides closed-loop monitoring and increases reliability. Another risk involves the use of a different planting method – transplantation, where baby lettuce is planted in greenhouses and then transplanted into the field. Will this work for lettuce? Lettuce is delicate, hard to transplant. This technique can place lettuce at the exact required position thus eliminating the need for thinning, but it would require a greenhouse. Transplantation could also be traumatic for the plant.

Has your business model shifted?

"Yes – before the model involved selling of machines, now it is a service model. It is working well – we charge per acre with a slight premium over manual labor. But scaling of a service business is a lot of work. We have three machines now, soon to increase to six. The logistics then will become more complicated and challenging. For the near future, we plan to stick with the service model – we will see what happens when we have 10, 20, or a hundred machines. We have been approached twice by a larger service company about taking over the service part of our business. Other crops are not thinned like lettuce but involve more weeding – we see great interest in weeding for diverse kinds of crops, to use less chemicals, and not spray all over the crop as insisted upon by environmental and consumer advocates."

Financing

The company started when both founders were students at Stanford with an idea, sweat equity, and no income for almost a year. A seed round of $175K was raised from personal savings, a friend, and two ex-professors, to buy much-required equipment. In 2012, Lee who had by then been earnestly developing computer vision models for over a year developed a good technical plan and submitted a proposal to the NSF SBIR program. They were successful in their first attempt and were awarded a NSF SBIR Phase I for $150K. A year later the startup succeeded in obtaining a Phase II NSF SBIR award for $500K. It was an important source of early funding and helped to establish credibility with private investors. Around the same time, the two founders met with approximately twenty venture capital (VC) firms, many of them based on the West Coast, with three specializing in agricultural investments based elsewhere.

Why look for venture capital and how much?

"When we started, we were building a robot machine with the focus on computer vision. It was meant to be a complete product unlike for example a software-based web application. It involved hardware, tractors, machines, and money to buy equipment. We looked at other options such as financing from early beta customers – they almost agreed, but it had drawbacks. It is a very competitive market, and money from one would have limited us from approaching other customers."

"The amount of capital required was determined by the best information we then had, and $5M seemed right. We talked to several other companies in the business to help us arrive at this number. In 2012, we were able to raise $3M from Khosla Ventures, negotiated a $1.5M debt instrument from Silicon Valley Bank (SVB), and obtained an additional $0.5M from angels. Having a lead investor like Khosla Ventures helped – they made introductions to SVB and helped recruit key people."

What about valuation and control issues?

Jorge answered - "I did my homework and had good people advising me. It is amazing how much of this is empirical, more art than science. It comes down to what you can negotiate. It is most important to have other alternatives. Khosla was fast out of the gate – we had their money, we had NSF money, and so at that point we could afford to be a little choosy. I could say thank you for your offer, let me try others, to get a better offer. Their rebuttal was that they couldn't guarantee this offer will remain. It's a game, an unscientific game of negotiation. We had many conversations before reaching agreement that both sides could live with."

"It is a painful experience. We did raise an additional $10M in 2014; Khosla participated in this round too, but another VC was the lead investor this time. This gives us a runway until at least the end of 2015. There are two board members – one representing Khosla Ventures and the other is me, one independent board seat is empty, and we have two observers. Both second-round investors are observers with no voting rights, but both have a say on who fills the third board seat. At the end of these two rounds the founders together hold a third of Blue River. The company plans a third and last round of VC funding in 2015."

Company Growth

Blue River has grown relatively rapidly. It now has twenty-seven people on the payroll, distributed among researchers, engineering staff, and off-site field personnel. Investors have been supportive of creating an equity pool for all employees. The Vice-President (VP) for Engineering, a computer vision expert, and the VP for business development are key personnel. The engineering team is responsible for the proper functioning of the current generation of the machine and is involved in developing future generations of the product.

Is hiring and retaining people with the right skills difficult?

"Hiring the right people is critical and hard and we have been lucky. It can take a frustrating amount of time to find the right person. We didn't compromise saying this person is "good enough" – no! For example, for our VP of engineering position we flew in ten people here to interview from all over the country. It helps that we are well-funded by brand-name VC firms. It's a long and elaborate interview process; we thoroughly research them before they come to the office. It's a full-day interview, and partly a working interview – if appropriate we ask that they write a bit of code. Each employee based on independent observations prepares a list of positives and negatives thus eliminating group bias. We have a good solid team that works well together. We spend a lot of time on interpersonal relationship building between new employees, founders, and early employees. We give them the company 'story' – share stories of what makes Blue River Blue River. Everyone tells stories. New employees hear this and get to learn company culture. We have peer reviews every six months. It is feedback from each other – what am I doing well, not doing well; people fill out a form that provides anonymous feedback; keeps us all honest. Consultants now and then come to talk to us on team dynamics. It is not easy with many smart people – headstrong opinions from seven PhDs for example. We have a team of stars. Our investors emphasize a diverse team that doesn't always arrive at the same conclusions on process, strategy, and technology development. It creates better outcomes but going through the process is not always easy."

Is having team members more "accomplished" than founders a problem?

"This is not an issue. It doesn't matter where ideas come from and from whom. The process of evaluation is unbiased. Being the founders, Lee and I constantly share the company vision with every single member of the team. We both believe sharing and believing in the company culture and vision are vital. You learn from sharing. Every two weeks we have hard-hitting thirty-minute sessions with the entire team - what's going well, what's not? We send out a survey beforehand for questions and issues people may have. We have biweekly $10 gift cards as a prize."

In terms of growth for the next 3-5 years, the company plans to fill out the lettuce business. There are currently three machines in full operational mode with one of them the latest generation. In the next two quarters the company plans to deploy an additional four machines in the field, with the goal a year from now to garner 10% of the U.S. lettuce market, moving up steadily in the next 3-5 years to capture 60-70% market share. The latest model machine can cover eighteen rows in a single pass while the previous generation machine covered eight rows. The biggest difference between the two versions is a built-in secondary vision system comprising double the number of cameras to provide auto-correction that minimizes mistakes. The machine is now capable of greatly enhanced levels of processing with no additional operator training required.

The first two machines have been successfully operating in the field for a year and the third machine for two months. Farmers think the new machine is a big step forward with innovative technology features but as always to them the proof is the actual performance in the field. In five years, the goal is to become a 100-person company with an engineering staff of twenty for lettuce and another twenty for phenotyping to diversify beyond lettuce and for growth. Blue River's lettuce market share will probably be capped at around $70M. The weeding market is a riskier unknown – both the weed set and the crop set - and it is much more seasonal. Exit possibilities include an initial public offering or a merger with a larger precision agriculture company.

Key Relationships

Do you need strategic partners, or would a good set of customers suffice?

"At this point I think the answer is no, regarding strategic partners. We build almost everything in-house, with standard parts for our machine outsourced. For these, we have a choice of machine shops, but we do like the ones we are working with. There's no need to go to another state or overseas; moreover, the agriculture industry is centered in California for these kinds of crops. The computer vision aspect, algorithms, the design of the machine, the assembly of about a hundred parts and final checks are all done in-house."

"We foresee continuing this for another five years. Salinas is good for manufacturing this kind of hardware because it is not high volume; easier to work with the local supply-chain. The NSF SBIR program was an immense help early on providing us credibility; NSF awards opened doors and people took us more seriously. For computer vision, we can get all the assistance we need at Stanford if required but we will need help in ag-bio and plant genetics. Many new innovations in this area first come to humans, then plants."

The Startup Experience

How would you describe your startup experience?

"Jorge and I started in January 2011, and it is going to be four years soon. There were many high and low moments with lots of grit required. Agriculture is not usually seen as a glamorous industry; many days spent getting the machine to work – on the ground, in the field, dirty and hard, lots of such days, but it has been very rewarding to see us execute our strategy. We have only been partly successful with investors about our vision. They always want a big market and want to make sure we bought into their vision of market size. At dinners with investors, I always say - part of what we are eating was produced by us, we helped make this food here. This is meaningful to me, that food is made efficiently, in an environmentally friendly manner. To me how food is made is important and that every plant counts."

Do you have the right people on board?

"We must make sure the people around us are the right people, ready to go on a journey, on an adventure – people make the biggest difference. It is hard to get this right, hard to judge people. Most often they themselves find out if they are a good fit or if they are not in tune with our team. It's easy to tell when a person jells as progress is very fast. We found a few very talented people, but they just didn't fit, and so they moved on. It takes a while to work with a new group, to get acquainted with the subtleties of how to work with our group, to figure out how decisions are made, who they should go to for help, how to bring up innovative ideas. I have learned a lot of lessons regarding people, they are still being learned. Relationships are important, even beyond making good decisions; individual relations make all the difference. Even when I make bad decisions, make mistakes, it still is OK provided the team trusts you, trust is key. I should have spent more time on recruiting; not my passion, almost must be dragged into it but it is so critical. Networking is not something I like to do. I have learned that resource timing and proper resource allocation are crucial, especially in a startup situation when resources are so constrained. If you spend too much time on one thing, say if lettuce becomes dominant, it will drain resources from other required areas. Areas to spend time on are often not clear. I didn't realize it initially, didn't appreciate this when we started out."

Do you consider yourself successful?

"Sometimes, in a contemplative mood, I think of our lettuce machine as art, a beautiful, artful machine we brought into this world, and that positively affected lives. In the future ten-twenty years from now, when sitting in a coffee shop conversing with a friend, I watch someone eating lettuce, and I can tell him, there's a good chance that I wrote the algorithm to make that lettuce better. I'll feel pride in work well done, in creating what for me is a beautiful piece of art. Talk to the customer – build something for them not for yourself."

Is agriculture a calling?

"Yes, it is, I do feel it is my calling! I really do! When deciding whether to start Blue River, an advisor suggested that I read the book 'The Monk and the Riddle'. I did, and it helped me put the right meaning to what I do. Make sure a person is doing whatever he or she is doing for the right reason. Creating food, people are intimate with food, it becomes part of them. Agriculture also has one of the largest effects by humans on the planet, changing Earth's landscape, for me it is exciting, it is directly correlated to people and to the planet."

Why did he say you read it?

"Well, he had started two companies, and had many ups and downs; one especially was very down. Reading the book helped him and me – if you are doing it for the right reasons, even very low points can be taken in stride. Try and change the world, make it better, even if it is very tiny, create something beautiful and new nobody has seen before; success or failure doesn't matter, it is more about trying, gaining specific skills, special insights; what I learned is that a startup cannot just be about money, about the bottom line; it is about starting a whole new journey."

Where's the startup now?

Acquired by John Deere, Inc. - Deere is bringing more robots to the farm. The maker of John Deere agricultural equipment said that it's acquiring robotics startup Blue River Technology for $305 million. The deal is expected to close in September. Based in Silicon Valley, Blue River makes "see-and-spray" robots that affix to tractors. They use computer vision to identify plants in the field in need of fertilizer, pesticides or other costly "inputs" used to manage crops. The robots are primarily used on lettuce, cotton and other specialty vegetables. As farmers spend on more on inputs, they tend to spend less on heavy equipment. By helping farmers be more efficient in areas like fertilizing, Deere can free up more money for investing in tractors. According to The Robot Report, 50 robotics and automation companies were acquired last year in deals totaling over $18.8 billion.

Blue River Technology raised about $31 million in venture funding from Pontifax Agtech, Data Collective, Innovation Endeavors, Khosla Ventures, and others. The company claims that its "precision farming" technology can save farmers up to 90 percent of the volume of chemicals they use with more traditional approaches. Phil Erlanger, a co-founder of Pontifax Agtech, told CNBC that the deal generated "attractive" returns for its investors but declined to give specific numbers.

"More efficiency and innovation all along the agricultural supply chain is necessary to feed the world with a population expected to reach 10 billion by 2050. Incumbents in the agricultural sector will increasingly snap up or partner with tech start-ups for innovative solutions."

Case Study Questions

1. Thoughts on using robotics/computer vision in agriculture
2. Assess commercial opportunity/potential and value proposition
3. Can this business be scaled beyond lettuce and how?
4. What role did the NSF SBIR Program play?
5. Finding the right people for agriculture industry applications - team building/dynamics
6. Identify risk elements (technical, team, market, finance); how to manage/mitigate them?
7. Start-up valuation exercise
8. Take-away(s)

2 Cleaning Water with Light

Technology Sector: Optoelectronics
Startup: Dot Metrics Technologies, Inc.
Website: www.dotmetricstech.com
Location: Charlotte, North Carolina

Federal Funding Timeframe: January 2008 – June 2014
Funding Amount: $1.24M

This narrative is a story of: M&A (multiple events: acquired by Aquionics, which in turn was acquired by Nikkiso); university as partner; joint development agreements; significant societal impact; scale-up/production challenges

Company and Team

Dot Metrics Technologies ("Dot Metrics") was founded in 2003 by Rosanna Stokes, an experienced new-product development manager. The company was formed to develop high-performance electronic materials and devices for the evolving solid-state lighting and photonics markets. Dot Metrics is based in the Charlotte Research Institute (CRI). The association with the University of North Carolina, Charlotte (UNCC) and CRI helps the firm minimize capital costs through agreements with the university and the Institute to lease equipment and rent space respectively. This allows the company at this stage to focus resources on developing critical technology rather than in building-up infrastructure. In addition to accessing state-of-the-art equipment, Dot Metrics also draws from a large breadth of technical expertise available at UNCC. The company is bringing to market a low-power, point-of-use water disinfection system.

Dot Metrics currently has five employees and two faculty partners at UNCC. Rosanna Stokes, Chief Executive Officer (CEO) and President, has over twenty years of industrial experience bringing new products to market. As a business development manager at General Electric (GE) Global Research, she worked with various GE business units identifying novel technologies to translate to commercial opportunities. At GE Silicones, she was Global Product Manager responsible for developing pricing and product strategy for products worth $600M. Jennifer Pagan, Director of Research, has ten years of microfabrication and rapid prototype development experience, and expertise in optoelectronics and wide-bandgap semiconductors. Dr. Pagan leads work on the development of novel optoelectronic device prototypes with a focus on design for manufacturability. She has served in lead technical roles in two other small businesses.

Paolo Batoni, Research Lead, is a GE Global Research Center (GRC) alumnus. He has extensive experience in ultraviolet (UV) emitters and detectors, and wide-bandgap semiconductors. Nadine Russel, Financial Manager, is a graduate of the University of North Carolina, Chapel Hill. Her background includes M&A experience, integration management, and international business. Clarke Monroe, Embedded Systems Engineer, combines his passion for innovation and creativity to optimize performance, energy consumption, reliability, size, and cost of the proprietary micro-controlled systems being developed by the company.

The UNCC faculty partners are Dr. James Oliver, Professor of Microbiology, and Dr. Edward Stokes, Associate Professor in Electrical and Computer Engineering. Jim has thirty-five years of experience studying bacteria from a variety of habitats and is considered an expert on environmental stress on bacteria. Ed is the developer of the proprietary platform technology used by Dot Metrics. He spent sixteen years at GE GRC where he most recently served as project leader in the Semiconductor Technology Lab developing white and ultraviolet light emitting diodes (LED) products for commercial and military applications. For several years, he worked closely with polymer chemists and chemical engineers at GE Plastics developing state-of-the-art instrumentation to enhance manufacturability of various thermoplastic products.

The Board of Advisors consists of Kenn Vest, Managing Director, Surfatas LLC, and Vice President of Marketing and Sales at Ricura Corporation. Kenn helps Dot Metrics with his successful entrepreneurial experience and understanding of the water treatment market. He is currently active in two companies marketing air/water filtration ingredients, technologies, and systems. Dr Stephen Lebouf, CEO of startup company Valencell, has filed over thirty patents on optoelectronic technologies and sensor systems for GE and Valencell. He led the successful development and field validation of GE Security's optoelectronic biological "smoke detector" for commercial and consumer applications.

Market Opportunity

As global populations expand rapidly and demands for limited fresh water supplies increase, *"water is the new oil"* has become a catchphrase. The global water market is $500B in size and growing fast. Dot Metrics is addressing the need for clean water with a low-cost, small-footprint point-of-use water disinfection system that companies are eager to test and ultimately license or purchase. The firm's proprietary UV LED disinfection flow cell in combination with current water filtration units addresses a critical customer need. Current UV disinfection technologies use mercury discharge lamps – these are bulky, fragile, require high voltage, and their disposal is environmentally problematic. The company can provide a unique product not currently available in the market by taking advantage of emerging low-power high-reliability semiconductor UV LED technology.

As municipalities have increased their reporting of U.S. drinking water quality to the public, the average consumer is becoming increasingly aware of large variations in water quality. For example, during one recent 27-month period, 24% of U.S. community water systems violated safe drinking water standards for microbes one or more times. These microbes indicate the possible presence of harmful bacteria, viruses, or parasites associated with human illnesses. The use of the major chemical disinfectant, chlorine, has been shown to generate a byproduct that has been linked to reproductive health issues. In the U.S., a relatively safe source of water, nearly 600 waterborne disease outbreaks have been reported in the past twenty years.

In addition to concerns about water safety, continued water shortages are providing a significant threat to global economies. The demand for water continues to escalate at unsustainable rates. Globally, water consumption is doubling every twenty years. By 2025, it is estimated that about one-third of the global population will not have access to adequate drinking water. Shortages are driving rapidly increasing prices for water. In the last five years, municipal water rates have increased significantly, by way of example in the U.S. by 27%, and in Canada by 50%. Reuse and point-of-use treatment of water offers one of the best methods for combating water shortages. It is estimated that the average household can reuse approximately 50% of their water. As scarcity drives up the cost of fresh water, more efficient use of water will play an increasingly significant role in water consumption.

Consumers are becoming more concerned about water quality issues and residential water treatment systems are experiencing significant growth. Fueling this growth is public awareness of the inadequacies of the global water supply. Globally, there are approximately five hundred companies selling residential water treatment systems. These customers are focused on the health of their family, and many can spend $500 on an easy-to-use and install, small-footprint water treatment system. The primary customers for the company's germicidal flow cell are suppliers of water treatment devices. These suppliers have the experience and ability to market to a vast array of end-users that require disinfection technology. The combination of a low-cost disinfection flow cell and a filtration system will offer a new product offering these suppliers can license or buy from Dot Metrics. This approach also provides this startup the ability to expand into other disinfection opportunities identified by such suppliers.

To obtain safe drinking water the health-conscious consumer is looking to obtain a source of water that is chemical-, virus- and bacteria-free. Dot Metrics' LED-based UV disinfection flow cell will meet customer's needs by providing disinfection through UV destruction of bacterial deoxyribonucleic acid (DNA), coupling with existing diverse filter elements to provide particulate removal using a small footprint, eliminate the need for chemical disinfectants such as antimicrobial agents, and provide battery operation for water and air disinfection. Initially, Dot Metrics plans to enter the U.S. water disinfection market. Once the company proves the technology, demonstrates its superior water disinfection capabilities, and educates potential customers on the benefits of chemical-free water, it will move into other global markets with the help of strategic partners.

Market drivers for UV LED disinfection technology are health concerns over disinfectant chemicals in water, global population growth and a continuing shortage of portable water, lower initial investment required for a water treatment system, and increasing need for portability and ease-of-use that can command a premium. This technology can offer battery or solar operation in situations where line voltage is not available or is cost prohibitive. LEDs are potentially four-to-five times more energy efficient than current UV bulb technology. This large energy savings can lead to use of batteries to power UV LED disinfection devices. For successful market entry, it is critical that the company's UV disinfection flow cell be qualified by the National Sanitation Foundation. Qualification requires considerable time, cost, and expertise, the last provided by their university partner, Jim Oliver. To obtain the required funding to qualify their product, Dot Metrics plans to partner with water filter suppliers and/or seek venture capital investment.

The U.S. market for water treatment will expand rapidly if Americans convert to drinking water from the tap than consume bottled water. In 2006, Americas bought almost thirty-five billion bottles of water. Three out of four Americans drink bottled water. It takes 1.5M barrels of oil per year to produce the plastic bottles for bottled water consumption in America. Bottled water has come under increasing scrutiny regarding their environmental impact - many large U.S. cities curtail bottled water usage. Bottled water drinkers are often under the impression that bottled water is safer.

Sixty percent of the usable fresh water in the world is in only ten countries. The limited freshwater resources and the imbalance of the location of their sources relative to population centers will continue to put pressure on global water supplies driving a need for low-cost disinfection technologies. The development of energy efficient UV LEDs for water sterilization purposes has the potential to provide a low-cost source of potable water for the world. If efficiency improvements using this technology are maximized, then it is likely that disinfection of water using small UV LED systems could be powered by solar cells.

Technology and Product

Dot Metrics is developing systems in the evolving market of UV LEDs. The company originally focused on LED development, but recent technological developments and commercial opportunities have redirected company resources on UV ED disinfection systems. Employing a unique systems approach, the company has developed a water disinfection system that is ten times more efficient. The system uses polychromatic UV radiation from UV LEDs to target multiple absorptive components within a pathogen[8]. The targeted radiation in conjunction with an efficient mechanical and optically designed reactor chamber provides a unique uniform flow, and a significantly enhanced pathogen inactivation cross-section. When used in combination with current water filtration units, this proprietary disinfection flow cell will fill critical needs for a variety of customers. One of the key advantages of this technology is the potential to use solar and/or battery operation in situations where grid electricity is not available or is cost prohibitive. The ability to provide on-site, no-power-required "safe drinking" water to disaster areas and needy world populations is significant.

This R&D project funded by the NSF SBIR program will bring to market a low-power, point-of-use water disinfection system designed to retrofit into existing passive (non-germicidal) filtration systems. This UV LED technology along with a novel, proprietary flow cell design improves upon conventional flow cells by maximizing the ultraviolet dose received by microorganisms in the water and increasing their residence time in the flow cell. Current UV point-of-use water disinfection is accomplished using discharge lamps that require high voltage, ballasts, and a relatively large form-factor. The use of UV LEDs will allow the light sources to reside inside a smaller form-factor, and to function at lower overall electrical power usage, without line voltage and ballasts.

The objective is to refine the current design to focus on manufacturability, and to test the prototype with Environmental Protection Agency (EPA) designated microorganisms. It is anticipated that the prototype will be as effective as commercially available discharge lamp flow cells while maintaining lower overall power consumption. The low-power aspect and small form-factor of the flow cell will make the system potentially suitable for battery operated field applications where line voltage is not available. Societal impact should be significant, particularly in markets outside the United States where there is increasing concern about water sterility.

Dot Metrics is currently developing novel flow cell technology for water disinfection for under-the-sink, point-of-use. However, there are a broad range of potential applications for the flow cell technology such as the opportunity to develop the flow cell for refrigerator water disinfection for a world-leading appliance manufacturer. This will create a prototype to fit within design constraints determined by this partner for a commercial product resulting in the first application of UV radiation in a residential refrigerator. This strategic partner manufactures over one million side-by-side freezer and bottom-mount freezer refrigerators every year for distribution throughout the world. Currently, there are very few refrigerators which offer microbiological claims for their drinking water as can be made using Dot Metrics' water disinfection flow cell technology.

The startup anticipates that its design can provide a superior total disinfection system for less than five hundred dollars. Significant costs for the consumer in any water treatment system are incurred in cartridge and filter replacement. Semiconductor LEDs are inherently long-lived, and the firm estimates that with current UV LED technology the UV disinfection flow cell will be replaced at every fifth filter replacement. The cost of treating a gallon of water using this system is comparable to low-cost filtration products which do not offer bacterial kill. The key to commercial success is to keep the cost of the UV disinfection flow cell to a minimum without adversely affecting its performance. The key contributor to cost is the UV LED itself but the company's proprietary design uses a minimum number of LEDs. Cost can be further reduced by purchasing bare LED chips at lower cost and integrating the LED packaging with the flow cell assembly. A primary focus of the NSF SBIR-funded research is to optimize the design to minimize the number of LEDs required.

Competing lower priced units do not offer any type of bacterial or virus kill. Current filtration technology traps bacteria but does not inactivate it. Without antimicrobial treatment of the filter which introduces chemicals, bacteria will multiply on filters and eventually break through. Filters have no effect on viruses. Chlorine and ozone treatment leave chemical byproducts in the treated water that could lead to human health concerns. UV mercury lamps and reverse osmosis require high power and a large footprint. UV LEDs are energy efficient, consume little power, and thus their ability to operate on solar or battery power can help to provision clean water to many needy world populations. UV mercury lamps are fragile and bulky. Reverse osmosis units are large and waste 5-7 gallons of water for every gallon of water treated. Although LEDs emit much lower doses of UV radiation, they can be operated in pulsed mode (turned on and off rapidly) thus extending the life of the UV LED - this operation is not possible with a mercury lamp. The optical design developed by the firm exploits the strengths of the UV LED in comparison to a mercury lamp while mitigating the downsides of the LED technology.

Dot Metrics and UNCC jointly filed a provisional patent application which covers their water disinfection device. An initial prior-art search revealed that while several patent applications and one issued patent describe using a UV LED for water disinfection, none employ a method like the optical flow cell design described in this patent application.

Business Model and Execution

The global residential water treatment market is growing at a compound annual growth rate of 19% primarily driven by Asia. Dot Metrics is partnering with filtration companies to bring the UV flow cell technology to market while at the same time not formally aligning with any specific company. The approach is to work informally with several water treatment suppliers. The UV flow cell is designed to be an add-on unit to current water filtration systems. The firm plans to first demonstrate feasibility in the U.S. market with partners and then quickly develop a global product. In developing countries where the need for water disinfection is the greatest, low cost is the critical driver. A battery or solar powered retrofit will allow the startup to also capture market share for current customers of simple filtration systems that are interested in replacing filters to achieve bacteria and virus removal.

There are numerous additional applications of this technology including air and surface disinfection, UV-curable adhesives/coatings, biological detection triggers, and medical applications requiring sterilization. The company has chosen to first pursue the water disinfection market due to the size of the market opportunity and significant consumer demand. It can produce the UV flow cell and/or license the technology depending on the market environment and partner needs. In 2011, Dot Metrics signed a Joint Development Agreement (JDA) with a world leader in UV technology supplying disinfection systems for municipalities as well as industrial applications to commercialize a water disinfection reactor based on their technology, and to develop a beta product for a select group of customers to test and validate this innovative approach to water disinfection. In addition to product development funding, this partner will help close key gaps by providing qualification testing, and critical marketing and sales support. Dot Metrics and its partner recently announced the commercialization of their first product.

Initially, the water reactor targets mercury-sensitive pharmaceutical and healthcare customers that are willing to pay a premium for a mercury-free solution currently not available. NSF SBIR funding will further development work using targeted wavelengths for enhanced inactivation of certain microbes and viruses. These funds will also help develop a disinfection cell with increased flow rates required for other market opportunities. Dot Metrics negotiated a JDA with a large manufacturer of refrigerators to integrate their flow cell technology into their products. This partner will provide detailed specifications for the specific part and the refrigerator model, prototyping expertise, and equipment such as rapid prototyping of flow cell parts, and prototype testing and manufacturing. Bacteriological testing will be performed at a certification facility such as those run by the National Sanitation Foundation. Successful development of the flow cell part will lead to its integration into 20,000 refrigerators in the first year.

The total NSF SBIR funding to Dot Metrics in the period 2008-2014 has been approximately $1.2M. Potential revenue streams consist of sales of UV disinfection flow cells and the sale of replacement UV disinfection flow cells. UV disinfection technology with LEDs rather than UV lamps being accepted as a reliable and qualified disinfection technology will mitigate risk and ensure successful market entry. Partnering with established filtration companies offering adaptations to their current product platform already addressing unmet customer needs at a reasonable price will further reduce market risk.

NSF Commercialization Assistance and Impact

In 2009, the NSF SBIR program launched a new initiative, the Innovation Accelerator (IA), the private component of a public-private partnership. The goal of this partnership was to facilitate the commercialization efforts of high-technology small businesses funded by this program. Dot Metrics was introduced to IA in 2010 by their NSF Program Director. The startup's needs as identified by IA converged upon the company's intellectual property (IP), successfully negotiating joint development agreements, and funding requirements. Contemplated issues include negotiating IP from the University of North Carolina at Charlotte, negotiating JDAs with multinational strategic partners, and providing potential investor introductions.

IA activities and outcomes are summarized in the following Table.

IA Activity	Outcomes
Startup Overall Engagement	34 months; 100+ hours of interaction; 220+ e-mails sent to and on behalf of Dot Metrics
Operational Assistance	Offered help to CEO in all operational areas
Negotiate Commercial Deals	Secured IP license from University of North Carolina, Charlotte
Introduce Potential Customers/Partners	SEMATECH
Find Investors	Southern Capitol Ventures; National Innovation Fund; Commonwealth VC; Siemens
Identify Domain Experts	AUTM; David Milligan, Advent IP
Feature at Trade Shows and Special Events	Angel Capital Association Conference; May 2012 NSF SBIR Phase II Conference Panelist

Testimonial

Dot Metrics' NSF SBIR Program Director note to Erik, a key employee of IA - "I want to congratulate you on this work. This company (Dot Metrics Technologies, Inc.) has made tremendous strides under your supervision."

Dot Metrics' Director of Research, Jennifer Pagan, was very complimentary of Erik's help in reviewing the two JDAs – "We recently had a significant new-product launch through a partnership and JDA with a world leader in water disinfection systems to market our UV-LED sterilization technology to the pharmaceutical industry for water purification. We also have a JDA with another large manufacturer of refrigerators."

Where was this startup at the end of federal government funding?

A joint development agreement (JDA) with Aquionics Inc. of Erlanger, KY was signed. The purpose of the JDA was to commercialize the UV-LED based water disinfection technology developed by Dot Metrics under the NSF Phase I and Phase II SBIR program. NSF supplemental funding was intended to bolster development of UV LED based lamps, or Uvinaire™, as they have been trademarked. The supplement was also used to develop a more efficient and smaller system per customer feedback.

Products:

1. The PearlAqua™ water disinfection system sold through Aquionics Inc. The link to the product page is http://www.aquionics.com/main/pearl-brand2/pearlaqua/

2. The Uvinaire™ is a UV LED lamp with integrated drive circuitry and is sold directly by Dot Metrics. The Uvinaire is the light engine for the PearlAqua.

3. The UV LED collimated beam system, a collimated beam for Petri level biological testing and dose evaluation, is also sold directly through Dot Metrics

Patents:

Patent application PCT/US2009/068765, titled "Systems and Methods for Performing the Bacterial Disinfection of a Fluid Using Point Radiation Sources" has been granted in Canada, Korea, and Japan, but is still pending in several other countries including the United States. Three other related patent applications have been filed and are listed below:

1. T. R. Harris, J. Pagan, P. Batoni, and J. R. Krause, "Apparatus for Irradiation," U.S. Patent Application: 14 202 969, Filing Date Dec 11, 2013.

2. P. Batoni, J.G. Pagan, J.R. Krause, E.B. Stokes, "Systems and Methods for Disabling Compressible or Non-Compressible Fluids Using Engineering Polychromatic Point Radiation Sources," U.S. Patent Application: 61 940 445, Filing Date Feb 16, 2014.

3. J. Pagan, E.B. Stokes, P. Batoni, "Optical Density Monitor and Comparator," US Patent Application: 61 820 204, Filing Date May 7, 2013.

Direct sales for the products listed total approximately $99,000. The number of employees at the start of federal funding was four, and five at the end of this funding.

Where's the startup now?

Currently UV treatment of fluids is almost exclusively carried out using low and medium pressure lamp technology that incorporates mercury as a generation source for UV photons. While UV treatment is replacing traditional chemical treatment in many applications, it still has many drawbacks. UV light-emitting diodes (UV-LEDs) provide solutions to most of these drawbacks. Dot Metrics engages in research and development aimed at integrating emerging UV LED technology in existing and future systems. Biological disinfection, as an industry, stands to benefit from the inevitable advances in deep-ultraviolet LEDs. Much of Dot Metric's current research is targeted towards this sector. Dot Metrics maintains a portfolio of intellectual property built around the unique characteristics of diode-based radiation emitters and the resulting applications. UV LED-based solutions for both research and OEM production are available.

Aquionics, Inc., the world leader in UV-C LED disinfection systems, acquired Dot Metrics in 2015. Aquionics is an award-winning manufacturer of water, air, and surface disinfection systems with UV-C LEDs (The UV-C portion represents wavelengths from 200 nanometers (nm) - 280 nm)) at the core of each design. It manufactures the PearlAqua, an award-winning UV-C LED water disinfection system; the solid-state technology used offers benefits previously unseen in UV disinfection. The company believes UV LED applications will improve health standards and boost the disinfection economy. Aquionics was in turn wholly acquired by Nikkiso America, Inc in a cash deal. With over a half-century of product development, Nikkiso has provided original technologies to a range of industries, including medical, aviation, life sciences, and microelectronics. Aquionics and Nikkiso share a common goal of providing life-saving disinfection products employing UV-C LEDs. Nikkiso is a leading manufacturer of UV-C LEDs, utilizing core technology from Nobel Prize recipients.

This acquisition allows Nikkiso and Aquionics to accelerate market commercialization of disinfection products around the world. "We have been excited to see the innovative solutions at Aquionics enter a number of high-value markets and look forward to supporting those efforts," said Dennis Martin, CEO of Nikkiso America. Commenting further, Oliver Lawal, CEO of Aquionics said, "We have worked closely with a number of suppliers to integrate the best UV-C LEDs, and Nikkiso has consistently delivered a strong product to us." Lawal will continue in his role, together with Jennifer Pagan as CTO, and all other employees.

Case Study Questions

1. Comment on the startup's market entry strategy
2. Is their business strategy/business model appropriate?
3. Strategic partnering – is this the best approach for the long-term?
4. Your thoughts on competition and barriers to entry
5. Was IA's help critical at this stage of the startup?
6. Can regulatory issues scuttle success?
7. Develop an alternate exit strategy for the company
8. Take-away(s)

3   *I Can Compute Emotion*

Technology Sector: Software
Startup: Affectiva, Inc.
Website: www.affectiva.com
Location: Waltham, Massachusetts

Federal Funding Timeframe: January 2011 – February 2016
Funding Amount: $1.28M

This narrative is a story of: The use of private equity funds beyond federal funding; multiple strategic partnerships; IP play; business model experimentation; multiple CEO changes with founder now the CEO

Startup Formation

Rana el Kaliouby, Ph.D. and Rosalind Picard, Sc.D., are the two co-founders of Affectiva, Inc. ("Affectiva"), a MIT Media Lab spin-off. Rana grew up in Egypt where she completed Bachelor and Master of Science degrees in Computer Science at the American University in Cairo. She found the subject of human-computer interactions more fascinating than areas such as operating systems, and computer design and architecture. Rana won the President's Cup for topping her class with a grade point average of 3.99 out of 4.0. Her career objective was to join the university's faculty - to accomplish this she decided to pursue a doctorate in computer science. In 1998, her husband who owned a software company in Egypt stumbled upon the book - Affective Computing by Rosalind Picard - on how technology can respond to human emotions, how computers can possess the ability to recognize, understand, and even to have and express emotions. He urged Rana to read it. She was inspired by its contents and the host of technical challenges it presented. Rana, by now particularly interested in the human face and its ability to convey a vast range of emotions, was accepted to the Doctor of Philosophy (Ph.D.) program at Cambridge University in September 2001 at the time of the 9/11 terrorist attacks. Her family was concerned about her safety and pleaded that she postpone going to England for at least a year. Rana was determined and would not agree to forsake such a life-affirming opportunity.

At Cambridge, she gave a presentation to the Computer Science Department explaining the many challenges attendant to making computers read emotions. An audience member shared that his brother was on the autism spectrum suggesting that researchers like Rana should find ways to use computers to help such children. This triggered many discussions with a professor who had for long been compiling data specific to emotion sets to help autistic kids with emotion-reading. Rana thought she could apply similar ideas and the professor's data to train computer algorithms to recognize facial emotions. In the summer of 2004, her role model Rosalind Picard visited Cambridge to give a keynote speech on affective computing. During her visit, Rosalind requested that she meet with research students working in this area. She had been following Rana's research and liked her approach – she offered her a postdoc position at MIT starting February 2006. Together the pair submitted a proposal to the National Science Foundation (NSF) to build 'wearable' social-emotion prostheses. Funding was denied - although the proposal ranked high on intellectual merit, it was deemed too risky and therefore not feasible. The two researchers called and convinced NSF that they had already developed various pieces of the technology platform, that they were engaged in building a prototype, and that their project was worthy of funding. Rosalind and Rana were subsequently able to obtain several rounds of research funding from NSF to pursue their ideas and refine their concepts at the MIT Media Lab. This Lab has an unusual structure for an academic department, with 80% of the work at the lab sponsored by industry and the rest supported by grants from federal agencies such as NSF and the National Institutes of Health (NIH).

"Sponsors visit the lab twice a year during Sponsor Week; and it is demo or die! We began to demonstrate our autism work in 2006. Sponsors like Samsung, Google, P&G, and Nokia saw an opportunity for our emotion-sensing technology beyond autism research, such as product testing and mobile phone applications. Over the next two years I began to compile a list of companies and use-cases that were of interest to them. In November 2008, Rosalind and I asked Frank Moss, the Director of the lab at the time, to assign us more researchers to fulfill the interest from sponsors. His response was that it is time to spin-out a company! Frank seeded this startup idea in my mind – and the tipping point for me was the notion that people can use my research. Why limit us to publications when this research can be used to help drive decisions in the real world. Rosalind and I formed our company in January 2009. We both had little business experience but decided to give it a try. The ecosystem at MIT is great. The Venture Mentoring Service (VMS) assigned us eleven mentors from different areas – law, business, investment funds – they repeatedly tore apart versions of our business plan and introduced us to the community of VMS companies to hear their stories."

Technology and Intellectual Property

In the minds of the two founders, their research and startup represent the marriage of two worlds – that of technology and humans, and business and academia. The core technology, Affdex, translates video data of people's faces to probabilistic emotion scores and helps to better understand human emotion using the principles of computer vision, image processing, machine learning, data analytics, and deep learning[9].

"Emotions drive behavior and 'emotion' data is of immense value to businesses, game developers, brand managers, and ad agencies. We capture this data unobtrusively and customize it for specific partners and their applications. In the 1970s researchers developed a manual for facial action coding and scoring system – a furrowed brow, a smile, a grimace – each was an action unit. It was laborious to walk frame-by-frame to code, and so we automated it. The barrier to entry, we think, is a combination of proprietary data and intuitive experience of how to program and train algorithms to understand facial wrinkles, shapes, and textures. There are fundamental differences attributed to factors such as race, culture, and age. One can say a smile is a smile but there are so many subtleties that we need to understand – social smiles, genuine versus politeness smiles; cultural norms drive people in Asia, in general, to be less negative about products. The face can be very telling, very revealing – for example Asians normally don't tell it like it is."

From the very beginning the two founders understood the value of building a strong IP portfolio. Affectiva has filed over sixty patents around their core emotion engine and what the company perceives as strategic use- cases of emotion analytics in market segments such as wearables, gaming, and media. The company has instituted a formal process with their law firm on IP strategy, which allows any team member in the firm to come forth with a patentable idea. Rana oversees and develops a list of important patentable ideas that are strategic, which she vets once a week with the law firm. She spends considerable time thinking about possible claims. In addition, the startup initially licensed five patents from the MIT Media Lab. At the Media Lab, sponsors can generally obtain licenses to all IP generated in the lab. In this specific instance only Affectiva has been allowed to sub-license the above five patents.

What are the key technology milestones to achieve in the next two years?

"First our aim is to capture everything about a human face, every nuance, every squint of the eye, every muscle movement – machines still find this hard to do. In addition, we want to tie emotions to behavior such as buying a product, calling a friend, clicking on a video. The intent all-along was to tie emotion to behavior but we didn't have enough data until now, and it took a while to understand how best to do this. I was also naïve about things working fine in the lab but would it in remote, rural America for example? It's important to be out in the field and obtain actual reactions, understand better how people emote. In the lab, with a small group, it's hard to get these kinds of insights. To properly read faces, lighting matters too – we must find ways to handle dark images just like how they figured out mobile phone usage in the sun."

Customer Discovery and Business Model

Over the last five years the company has gained considerable knowledge of the $32B consumer-insights market, a market focused on understanding consumer responses to advertising, online video and products. This space has evolved – when the company started, people were cynical, and thought the technology too research-oriented; now the conversation with potential customers has shifted from the "why" to how they can integrate emotion analytics into their business processes. An important early adopter of the technology was Wire and Plastics Products (WPP), one of the world's largest advertising and marketing conglomerates. Also, a few Media Lab sponsors were the firm's first clients, for example testing different men grooming products for Proctor & Gamble (P&G). In most of Affectiva's customer discovery calls, there were innovators and internal champions in large companies who were willing to risk their careers on the successful adoption of its technology. It was critical to find those champions and help them be successful within their organizations.

"At the start, we did everything that landed in our lap. The P&G example wasn't a perfect fit for our technology. Soon we compiled a set of criteria that defined product-market fit, which we used to vet potential customers: emotion invoking, scalable, and repeatable. In the initial stages, we were particularly good at analyzing humor although our technology could cover a vast range of emotions. The NSF SBIR program allowed us to further explore some of the R&D challenges, in addition to providing equity-free funds. We also found that being grantees of this program greatly increased our credibility when we were out raising capital."

Affectiva contemplated whether they should go direct to brands or partner with market research firms that already have connections with brands and advertisers. The problem they encountered was that someone like Coke's brand manager most likely gets hundreds of calls from startups. To counter this, Affectiva decided to partner with Millward Brown (MB), a WPP company; MB essentially became a channel partner. In addition, the team learnt that within big brands there is always an innovation function. In fact, the company could engage P&G through its innovation officer, who was more receptive and venturesome to the latest technologies.

"Also, we find it hard to get people to directly sign on to a $1M project; so, we start small and do a pilot. Test six of your ads, we say, if you like the results then we can scale. Clients usually want validation work done first before signing a multi-year agreement."

How would you describe your value proposition and business model?

"We are on the cusp of an emotion economy, where brands are placing a premium on emotionally engaged consumers. Increasingly more and more brands want to build a strong emotion connection with their consumers, and they realize that this is the path to loyalty and purchase behavior, but they struggle to measure that in a scalable, scientifically valid way. Enter Affectiva's emotion-sensing technology and analytics! It's essentially a software-as-a-service model (SaaS). Partners sign up for a monthly or yearly subscription to use our platform and access our emotion norms. It is a volume-based subscription model. We started with a web-based portal but now we find that a lot of our customers just want the emotion data. They want this data fed into other platforms that they can derive value from analyzing in the manner they choose. We also have a mobile and desktop software development kit (SDK) which we sell as a license to app developers."

What about competition?

"The biggest competition is the status quo - the inertia of change in large organizations. Companies have used market surveys forever. It has worked for them, so why change? There are two other startups involved in facial recognition and emotion in voice. Although I believe we have a stronger science base, I have a great deal of respect for these other companies. We are all creating and that is what spurs innovation. At the end of the day, there is plenty of opportunity for more than one of us."

What made you apply to the NSF SBIR program?

"Our business plan involved developing a cloud-enabled platform for the analysis of emotional and cognitive states from the face, a commercial version of the MIT FaceSense, an NSF-funded, vision-based computational system that reads states such as liking and confusion from facial-video using any webcam. We were referred to the NSFSBIR program by two recent Media Lab startups that had funding from this program – Sifteo and Bluefin, the latter acquired by Twitter. Both were generous sharing their process knowledge of NSF SBIR. NSF grant money totaled approximately $1.1M – non-dilutive funding, which is great!"

Financing

In May 2009, Rosalind, and Rana each invested $150,000 in this fledgling venture to hire their first three employees – a hardware engineer, a software engineer, and a project manager. These initial funds were enough to get them through to December. The two founders did not pay themselves during this period. They started to raise funds in April 2009 and closed their first venture capital (VC) round of $2M in January 2010 with plenty of assistance from friends, advisors and mentors who helped them determine a fair deal. They had three term-sheets - one from the Wallenberg Foundation that had been following their research for several years, a second less favorable one from a top-tier investor in the U.S. and a third from a Middle East-based VC firm. The founders picked Wallenberg because beyond the money they shared similar values. The $2M gave them a runway of about two years. In April 2010, they hired a CEO, Dave Berman, a seasoned business executive recommended by one of the VC firms that was courting Affectiva. Berman started to build-out the rest of the management team. The burn rate started to rapidly increase forcing them to fundraise again. They closed Series B quickly thereafter with $7M from WPP.

Why raise so much money?

"We rapidly learned that there were many use-cases for our technology and many verticals we could play in. To build towards our larger vision of being the emotion layer in the digital world, we needed to raise additional money, so we could continue to enhance our cutting- edge science and technology, but also invest in outstanding sales and marketing talent to execute. So, we completed our Series C round raising an additional $12M from Kleiner Perkins and Horizon Ventures."

Company Growth

In the early days, both founders were heavily involved in day-to-day operations, investor meetings, and team building. Rana remains intimately involved in the company daily where she leads the 'Emotion Science' team. The original vision of leveraging emotion analytics to improve business and people's lives has not changed. An unfolding journey with many stops, the company is currently focused on media testing and the advertising space – it is not the destination, but a market where it sees immediate revenue opportunity. The Board consists of the Wallenberg Foundation, WPP, Myrian Capital, Horizon Ventures, CTO and Chief Science Officer Rana, and CEO Nick Langeveld.

What are the keys to success?

"I would first categorize it initially in the areas of science, market, and innovation - the core science team is critical; it is our competitive advantage. We seek out researchers in the field of affective computing and recruit them to join our team; finding the right product-market fit is critical - what keeps me up at night is that were so focused on use-cases in front of us that we may be missing the bigger picture. Now it is all about advertising and media analytics, but I believe there is a bigger application of emotion-sensing on mobile devices. One of my responsibilities is to find the next set of breakout use-cases as I better understand this technology's potential and its many distinct aspects and use-cases. I have been living with it for years. Admittedly there is a tension between exploring new markets versus focusing on tried-and-true revenue generation channels; and commitment to innovation: I have set up a process where we set aside a few hours a week with a small team to brainstorm the next use-cases and innovation ideas. Recently, we discussed the 2015 plan, and I deliberately took a backseat. The team collaboratively came up with three exciting new initiatives that others can take a lead on. I believe in intrinsic motivation; people do a better job if work is not inflicted on them. We have recently started company retreats and may even adopt the Google model (a day each week to work on one's own ideas) to keep everyone excited and passionate about how our platform technology can be usefully deployed in many different markets."

What then are the other categories?

"What I call culture and people challenges; a sense of ownership and empowerment: All fulltime employees, even if we grow to a 100-person company, have, and will have equity in Affectiva. In a small team, each one can make a difference, there's a lot of trust and empowerment, everything each does affects the product. The culture fit is critical - at one point we hired a team member who lacked technical depth and had a militaristic leadership style that did not fit with our culture of collaboration. I invested time and energy to get him up to speed, in his success, so it was not easy to let him go. When our first CEO left, I worried that this would have negative consequences for the company. But with a well-thought-out transition plan, we promoted Nick Langeveld, our VP of business development to CEO. It is working out well so far."

What would you like to see happen in the next 3-5 years?

"Our core technology can be transformative in at least three areas – market research, healthcare, and learning - but it is hard for one company to do all three. One big lesson learned is that we underestimated what it takes to succeed even in a single market. Maybe in five years we can have three entities separately addressing these three markets. The technology is new, trailblazing, with many possible use-cases to show the world the many possibilities. In terms of market timing, it's probably harder to tackle online learning at this point, but we are on the lookout for opportunities in this space. Potential partners understand what they want to do with our technology at a high-level but look to us with the expertise to help bring other applications to life. There is considerable interest from Fortune 10 companies, and in small companies, about using emotion sensing in the consumer wearable space. They perceive us to be a market leader in the use of computers to understand emotions that translate to human behavior. How we interacted with devices was different before we had touch-enabled gadgets. We'd like to look back five-ten years from now and ask – what was it like before such devices had emotions?"

Do you have an exit strategy?

"Organic growth to a 100-person company, M&A, IPO – my thinking on this is in flux. Maybe we will have multiple entities with Affectiva a holding company, maybe it is more efficient and effective to partner with a large company, with massive scale and reach, like an Apple, Facebook, or Google, with emotion-driven smartphones and other mobile devices. I am not sure – all I know is that I have a lot of things to get done before I stop."

Key Relationships

Rana considers the NSF SBIR program a key relationship that helped develop the technology and provided equity-free financing. It helped raise capital by lending additional credibility to the science and the team because of its peer review process and its highly competitive nature. The program director provided valuable initial advice regarding market course corrections. Prior to startup formation NSF granted multiple research awards to develop the core science.

"NSF Program Directors collectively have a lot of experience and can help startups such as mine even more in terms of sharing lessons learned, connecting us to other companies, helping with business models, and mentoring. It's never just about money but also about networking advice and being smart about choices. Grantee workshops are a nice feature of the program – I especially like the Phase IIB (supplemental matching funds to NSF SBIR Phase II grantees) presentations; these startups are further along and have valuable lessons learned to share with others. I keep in touch with two wearable-technology companies I met at this event. The women entrepreneur's breakfast meeting was valuable too. We continue to keep in touch, share learnings and experience from the perspectives of women."

How can the program do better?

"It'd be great if the awards recommendation process could be expedited. It sometimes feels like throwing a proposal into a black hole for 4-5 months – we are not told what's happening. It will help if we are able to track where our proposal is in the process. For example, if we'd heard that our Phase II was recommended for award earlier, we could have delayed our fundraising until critical milestones were met and technical risk further reduced. This would have increased our valuation thus lessening equity handed to investors. I do realize that NSF's review process is in-depth and thorough, and this takes time – I wouldn't want the quality affected."

The Startup Experience

Can you describe a high- and low-point in your startup experience?

"One highpoint: This was during our series B fund-raising – we did a pilot for WPP as part of their due diligence. It was a critical project for us. All hands were on deck for weeks, the schedule was tight, and we simply had to pull-off this project successfully. I was amazed at the energy, dedication, and true startup spirit of our team, all rowing in the same direction, all focused on the same goal. We passed the test and WPP decided to lead our Series B round of $7M!"

"The low point was when our first CEO left. I was concerned that with his leaving the rest of the leadership team might leave with him. We did lose some momentum, but I felt an increased responsibility to energize the team, be positive, honest, and authentic."

What did you learn about yourself from this experience? Did luck play a part?

"I am hard-working, focused, committed to success, persistent, will not take no for an answer, and will find a way around problems. I find it hard to switch off work, and often work late after I put my children to bed but must learn to bring more balance. I try to stop when I am with my kids to focus on them. I think it is harder to do a startup being a woman, a mother, and an immigrant, but I have been fortunate to have a good support network of people around me."

"I am passionate about our technology and believe that somebody will impact society in a positive way using this technology, and I want to be that person. Generally, I think luck has not much to do with success although chance encounters are a kind of luck – I stumbled on Rosalind's book, a person in the audience for a talk I gave brought up autism, Rosalind visited Cambridge and offered me a post-doc position at MIT; opportunity presents itself and one has to be prepared to take it."

What advice would you have for a would-be entrepreneur?

"I'd encourage them to go for it but to be well-prepared before they jump in – acquire as much knowledge about the technology, and the market as they can. They should be networked to people of all kinds – researchers, investors, mentors, and other companies and startups. Are they passionate about the idea? Does it address an important problem or need, is it a cause worthy of pursuing? Don't pay too much heed to naysayers who say this has been done before, that nobody will fund you – don't be fazed by negative comments."

NSF Commercialization Assistance and Impact

This case study provides an example of the Innovation Accelerator (IA) deploying resources - time and people - in a seven-month effort to assist a startup in its very early days with no tangible directly related outcomes to show for their work. It points to a case where the surrounding innovation ecosystem (MIT in Cambridge, Massachusetts) makes available to a startup numerous support mechanisms and infrastructure that obviates the need, in some cases, for such assistance provided by the NSF SBIR program.

IA and Affectiva were introduced at the NSF Phase I Grantee Workshop in March 2011 upon the recommendation of their NSF Program Director. Through initial conversations with the startup team, IA first tried to understand the technology - to develop the Face Reader Platform, a cloud-enabled SaaS platform for the analysis of emotional and cognitive states from the face. This platform addresses potentially lucrative business opportunity in market research, media research, product testing, and usability testing, offering insights into customer resonance. The approach consists of building a multi-tier architecture that makes facial expression analysis seamless, scalable, and affordable.

IA then performed a needs analysis – it engaged Affectiva's CEO Nick Langeveld, who had a strong business and operational background. It found the startup to be more mature as a company than any other Phase I or even most Phase II companies IA had worked with. IA's assistance was required in the following areas - introductions to VCs for their Series B round; arrange meetings with Interpublic Group (IPG), a global leader in modern marketing solutions; help with customer development; and identify talent to build-out the initial team. This activity involved seven months of engagement, over fifty hours of interaction with the startup, and sending/responding to over two hundred emails to or on behalf of Affectiva. The company was introduced to Fairhaven Capital, InQTel, NorthBridge Venture Partners, GGV Capital, Commonwealth Capital Partners, Raptor Accelerator, and Chart Ventures. Introductions were made to domain experts Doug Billman, Mike Sigal, and Rob Phythian. Interactions with potential customers such as Facebook, IPG, Sports Media Advisors, and Silver Chalice were facilitated. Glenborn Associates and The Ladders were engaged to help add talent to the Affectiva team.

How did IA help?

"They were most helpful during our Phase I - they connected us with potential investors for our Series B round in 2011. They facilitated introductions and meetings with VCs mostly based in Boston, New York, and Washington D.C. We explained our technology although most of the VCs we met through them were more interested in investing in cyber-security startups. It still was valuable experience because we learned to speak their (VC) language. They also introduced us to market domain experts – that was useful too. I think we were not able to fully leverage their capability. I guess it's because the local MIT network introduced us to many potential VCs, and I do understand that other grantees don't have MIT-type networks. We did not build a close relationship with IA but it's not their fault."

IA effectively assisted Affectiva over a short period of time by directly addressing the company's needs. Even though the company chose to accept Series B capital from WPP, a strategic partner, IA's introductions to VCs in its own network created significant learning and leverage. IA's impact in working the back end of the IPG deal will be extremely difficult to measure; however, the relationship provided the startup with the assistance they needed. In terms of customer development, IA was able to provide introductions to well-connected firms in the sports digital media space: Silver Chalice whose Chairman, Jerry Rheinsdorf, owns and manages digital assets of many sports teams and brands, and Sports Media Advisors made up of former American Broadcasting Corporation (ABC)/National Hockey League (NHL) Chief Operating Officer and senior vice-presidents, who advise many companies inside and outside of sports on the digital media landscape including Goldman Sachs, HBO, NASCAR, and the National Football League (NFL). IA also engaged the Head of Marketing for Facebook, who eventually passed on the opportunity. In addition, IA managed to forge a relationship for Affectiva with The Ladders, one of the top executive search firms in the industry.

Where was this startup at the end of federal government funding?

In 2015, Affectiva secured several strategic partnerships: With the explosive growth of mobile video communications, massive amounts of data are generated, yet until now not fully leveraged to enrich the user's mobile video communication experience. Its first partnership targets one of the most fundamental elements of effective communication — emotional response. Through integrating Affectiva's Software Development Kit (SDK), the new platform measures emotion. To deliver a unique and highly engaging experience founded on a diverse set of user insights, this partner has developed a platform it calls Intelligent Video™. A multi-year licensing agreement is in place, which includes royalties and revenue share.

The second partnership involves disrupting the hiring space through video-based recruitment. With Affectiva's technology, it is possible for this second partner to quantify and sort candidates based on their social aptitude. In addition, the same platform can be used as an interview coaching, or sales coaching tool for consumers. This is a multi-year licensing agreement. The third partnership involves human behavior researchers who can combine facial coding and emotion analytics with eye tracking, brainwave measurement (electroencephalogram (EEG)), as well as physiological sensors (galvanic skin response (GSR), electrocardiogram (ECG), electromyography (EMG)) on top of traditional surveys and questionnaires, which are also fully integrated. This enables them to, for example, assess the impact of new digital and learning experiences such as gaming, on affect and cognition and ultimately user behavior.

Where's the startup now?

Affectiva is a 2009 spin-out from MIT's Media Lab, and is backed by leading investors including Kleiner Perkins Caufield Byers, Horizon Ventures and WPP. Emotions influence all aspects of our lives, but are missing from our increasingly digital worlds, hampering how we communicate and interact. Technology and devices are hyperconnected and super-smart; they have high IQ but no emotion intelligence (EQ). By bringing emotional intelligence to our devices and digital experiences, we can close the empathy gap to foster better ways for humans to connect with each other and build more likeable and persuasive machines that can bring positive change to our lives. This technology can show how facial expression technology, and emotion sensing and analytics can transform our digital world. Following is a sample list of articles appearing recently in the popular press:

- New Affectiva cloud Application Programming Interface (API) helps machines understand emotions in human speech
- MIT Spinout Affectiva Adds Voice Analysis to Its Emotion-Sensing Tech
- Computers get emotional
- Emotion AI
- MIT-born firm aims to raise $20M to expand 'artificial emotional intelligence' tech
- The Case for Emotionally Intelligent AI
- Affectiva's AI Algorithms Can Tell the Difference Between a Smirk and a Smile
- Happy or sad? Your future car might know the difference
- Will ads be able to read your mind?
- Artificial intelligence that can read your emotions
- What emotion AI can teach us about human behavior
- Building Emotionally Aware Cars on the Path to Full Autonomy
- Can this Radio Detect your Mood?

Case Study Questions

1. Assess commercial opportunity/potential and value proposition
2. Business strategy/business model to adopt?

3. Team building/dynamics
4. Identify risk elements (technical, team, market, finance); how to manage/mitigate them?
5. Barriers to entry
6. Discuss multiple other potential markets
7. Growth and exit strategy
8. Conduct a startup valuation exercise
9. How best to align with venture capital expectations?
10. Take-away(s)

4 An Even Smaller Memory Chip

Technology Sector: Semiconductors
Startup: Zeno Semiconductors, Inc.
Website: www.zenosemi.com
Location: San Jose, California

Federal Funding Timeframe: January 2011 – May 2016
Funding Amount: $0.99M

This narrative is a story of: Industrial strategic partners; contracts, licensing, and joint development agreements; IP play (over fifty patents); scale-up/production challenges

Startup Formation

Yuniarto Widjaja is the CEO and sole founder of Zeno Semiconductors, Inc. ("Zeno"), a company he incorporated in 2007. Yuniarto migrated from Indonesia and arrived in America to pursue an undergraduate degree in chemical engineering at Purdue University, after which he received a Ph.D. from Stanford University ("'Stanford") in chemical engineering with a minor in electrical engineering. In Indonesia, Yuniarto's family owns a company that manufactures products for the consumer market. His first exposure to entrepreneurship therefore was from his own family business – to this day his role models remain his parents.

While at Stanford, 1997-2002, he was immersed in a vibrant culture of entrepreneurship. Many compatriots were eager to start their own companies or join a startup. At Stanford, there are innumerable entrepreneurial courses, business plan competitions, and seminars featuring young entrepreneurs who shared with the audience the mistakes they made, and the lessons learned from these. They made a solid impression on Yuniarto. The message he took away – you learn more from making mistakes. He spent countless hours with classmates and friends discussing possible ideas to seed a startup.

On the day of the first telephone conversation with Zeno, the team was busy with 'tape-out' scheduled for the next day. Tape-out is the result of the integrated circuit (IC) design cycle when the artwork for the photomask of a circuit is sent for manufacture. A photomask is a pattern layer in IC fabrication fed into a photolithography stepper or scanner, and individually selected for exposure.

Yuniarto, you started Zeno five years after you left Stanford?

"Yes, I thought it best to first get relevant industry experience in memory devices; get to learn the business, the issues and gaps in technology and customer pain-points."

Did you simply quit – how did it happen?

"I took a week off to think about it; intensely. Discuss it with my wife; needed family support and my wife's feedback. She was not against it; was encouraging and supportive. I set my mind to it, I decided to quit."

Was she working too – was there a second paycheck?

"No, in fact we just had had a baby! You know, there is no appropriate time to start a company, you just start."

Technology

How did you decide on the technology to pursue?

"After Stanford, I worked for five years at Silicon Storage Technology mostly involved in the design and manufacture of memory devices, and specifically in flash memory. It was a valuable experience. I have some idea of the problems facing the semiconductor industry in general, one of these is in the memory space; it's also close to my background – both in industry and in my academic training. I knew of specific problems, so I thought hard on potential solutions, so this wasn't the case of a technology looking for a problem but the other way around – it was a problem looking for a solution. The technology piece was clear in my mind."

Although Yuniarto's Ph.D. work was not specifically geared to memory devices, he gained many insights in this technology area from five years of industry work. The physics involved in his Ph.D. work could very well be applied to the problem at hand. Moreover, he could count on technology experts at Stanford such as Professor Yoshio Nishi, the Director of Research, Stanford Center of Integrated Systems[10]. Professor Nishi has decades of experience in memory devices and has worked in companies such as Texas Instruments (TI), Hewlett-Packard (HP), and Toshiba. Yuniarto values his opinions and expertise.

Was the technology being developed from your mind alone?

"I had many discussions with friends and colleagues, learnings, and knowledge from Stanford. I also read many publications of others' work in this area."

So, what is the problem that you had identified?

"Well, one of the challenges with the on-chip memory problem is the real estate issue wherein more and more of the chip area is taken up by memory, now more than 50%. On-chip memory is mostly implemented with 6-transistor static random-access memory (SRAM). In our case, we are only using a single transistor."

This will lead to savings?

"Yes, cost savings, it can reduce the chip size or offer higher memory density. With the usual 6-transistor configuration we can create a feedback loop with four transistors forming two inverters. The challenge was how to create two-stable states with just one transistor?"

Where will the feedback then come from?

"Even within a single transistor there are many different devices – diodes, bi-polar transistors, et cetera that we can use; we employ the physics of such devices to obtain two stable states."

Zeno is involved in the development and optimization of a novel one-transistor memory device, and a memory-IP block that would have many possible applications – e.g., to enable power-efficient computing applications for mobile devices; to reduce power consumption in data centers; to provide an integrated memory solution by combining different types of memory used in typical electronic devices into a single memory device that combines their various specific characteristics.

Zeno now owns thirty issued U.S. patents. There was also no need to license technology from Stanford.

Did you need help with filing so many patents?

"We have a patent attorney; to help with writing patents and filing them with the patent office. Initially the attorney wrote the claims. Now we have several team members who are very experienced with intellectual property (IP). Therefore, the later filings, we were a lot more involved in defining the claims."

What's your IP strategy – why so many patents and the money needed to file them?

"The licensing model, that's why; moreover, there are so many new revelations that come up with the one-transistor concept – memory cell, circuitry; lots of room for inventions and innovations. To have a large patent portfolio is a useful differentiator for a startup company."

Would you need additional IP?

"Right now, we are still at the technology development stage with constant innovations ongoing – hopefully these will lead to more IP; more importantly, we have critical milestones to achieve in the next twelve months."

Customer Discovery and Business Model

At the beginning were theory and simulations, then a proof-of-concept device was developed at the Stanford Nanofabrication Facility and later with funds from the National Science Foundation (NSF) Small Business Innovation Program (SBIR) program Zeno fabricated a prototype that was used to find early partners.

"Initially, our technology development effort was kept under wraps; we deliberately avoided publicity, kept a low profile. With a ready prototype, we approached potential early adopters individually – they were intrigued but were hesitant to take the lead, they all wanted to be 'the first follower'; generally, technology marketing was not a challenge – lots of companies appeared interested; they all said, if this technology works, we'll use it."

The key early adopter was a major fabless semiconductor company ("Company A"), a technology-driven company with operations worldwide. This company is always enthusiastic about the latest technology; it is in their DNA. Zeno now has an agreement with this company that incorporates an investment component that is milestones-driven and a joint technology development component. Zeno obtained help with negotiating terms from their law firm whose attorneys have experience in negotiating memory licensing agreements. Company A was offered preferred terms because they are an early adopter. The agreement does not preclude Zeno from engaging other future customers.

This strategic partner understands the value proposition: moving from six- to one-transistor designs would result in smaller die size, lower manufacturing costs, and higher memory density. In the semiconductor supply-chain Zeno would be a solutions provider to fabless semiconductor companies like Company A and companies that have their own foundries around the world.

What is your addressable market?

"Our technology can be embedded in all kinds of chips, chips used by companies such as Intel, Samsung, Qualcomm, and Apple. However, right now we are focused on achieving the next milestone with our partners."

"Our technology can be embedded in all kinds of chips, chips used by companies such as Intel, Samsung, Qualcomm, and Apple. However, right now we are focused on achieving the next milestone with our partners because most companies will ask for data and performance specs, which as of now we don't yet have a complete package. That's why these partnering arrangements are important because we lack sufficient resources to prove out this technology on our own."

What then is your business model?

"In the beginning, I thought we'd be a product company, all the way to manufacturing the product by ourselves. As time went by, and we gathered additional market insights this began to shift to a licensing model. In the semiconductor business, the technology roadmap to build a product versus a licensing model requires a fundamentally different emphasis and focus – the scale and investments required for a product company can easily run into hundreds of millions of dollars. It also makes more sense for us to make it technology- and/or product-agnostic. Almost all electronic devices have active memory and memory blocks – microcontrollers, network devices, embedded processors, this is what we plan to license - whole memory blocks to place inside their chips. We can license to fabless companies and directly to foundries or even to memory device customers (large companies such as Intel) although the 'sale' in each case is different and complicated."

Can you comment on the competition?

"Many other memory startups exist because as we've discussed, memory is significant in the semiconductor industry; what's unique about Zeno is that we use standard silicon complementary metal-oxide semiconductor (CMOS) processes while many others use different materials. New materials are always very difficult to incorporate in existing manufacturing processes – we think this is a big barrier; there's always resistance even in the case of silicon to minor changes in processes or process technologies."

How do you assess risk?

"There's always the technology risk – we appreciate the NSF SBIR program that was key to reducing risks up to proof-of-concept and prototype; soon technology risk will be further reduced with Company A's help, and NSF too. As for market risk, the demand for memory is always there. When we discuss our technology, other companies all say – if your technology is proven, we will use it; the risk attached to adopting new technology exists. The question remains - how to bring to market quickly and continually innovate. Lots of protectable IP helps. In terms of financial resources, we can continue for another 2-3 years. I think it will all come down to execution and people – people risk."

Financing

After he had quit working at Silicon Storage, Yuniarto started Zeno with seed funding he obtained from his family. He was fortunate to also obtain additional funding from an angel investor. Together these initial funds amounted to approximately $200K. This provided a runway of 12-18 months for technology development. Around this time, the company did seek venture capital (VC) funding, but VC firms were reluctant to invest in a semiconductor startup this early in the technology development cycle. The technical risks were simply too high. Moreover 2008-09 was not a particularly conducive time to raise capital from the private equity market.

During this period, he had heard about the NSF SBIR program from friends and colleagues at Stanford. He decided to further research it and liked the fact that the topics covered by this program were quite broad. The only other SBIR program he considered was the one at the Department of Defense but soon determined that the topics of interest to him in that program were too narrowly defined; moreover, they were all contracts, not grants as is the case for the NSF SBIR program. Zeno was not recommended for funding the first time it submitted a proposal to NSF. It did however obtain quick and valuable feedback from the review panel, both on the technical, and on the business side. Yuniarto and his team of two other technologists realized that they had failed to articulate their technology innovation and its market potential properly and effectively. They were determined to try again, and this time submitted a vastly improved version of their original proposal. This time Zeno was successful and was awarded Phase I funds of $150K.

Did you have problems working with a federal agency like NSF?

"No, not really, but there was one aspect of the program that I'd like to share - when we were not funded in our first attempt, it would have helped to have more interactions with program staff – if they could have shared more insights with us as to why the proposal was not recommended and how best to improve."

Zeno applied for Phase II funding twelve months after award of Phase I when they were already eligible for such funding six months after the start of their Phase I. This was a conscious decision - because a startup can apply for Phase II funding only once, they strongly felt that their chances would be further enhanced if they had in place a solid relationship with a strategic partner or two. Such a partnership with Company A was recently cemented after Zeno was awarded the Phase II of approximately $493K from NSF. As a reflection of their interest and faith in the startup's technology, Company A has decided to invest many millions of dollars in Zeno. This investment is not meant to be over a fixed timeframe, it is instead based on technology development milestones. Company A has already made available initial funds in addition to co-development and in-kind resource sharing in terms of equipment and test facilities.

"The agreement with Company A took longer than expected – we started initial discussions a year ago, but the agreement was only signed early this year; it seemed long for us, but experienced people say it was relatively quick for a large company such as Company A. In the agreement, there are intangibles that are as valuable to us as the money and resources. For example, a high-value intangible is making foundries pay heed to our technology - if we approached them directly, they would say we are too small to pay attention, but if the request came from Company A, they'd immediately get to work."

Company Growth

Yuniarto is the sole founder of Zeno. Now there are three fulltime employees, all focused on technology development. Now he thinks they need a few more to help but has found it difficult to find people with the right skills –

"Fifteen years ago, people really wanted to join startups, maybe they still do but only with software companies. We are fine for now – in the next six months we do need at least one more, maybe two will be good, to accelerate the technology development. All three employees, Company A, our law firm that provides discounted services, technology advisors, and the angel investor have an equity stake in the company. There is also some equity stock set aside for future employees."

What if a key person leaves?

"That will hurt – if one leaves, it will be one-third of Zeno in human terms. This is a risk, and we should work more on this; maybe we've been putting it off because we are all so busy but later as we grow, I am sure people issues will spring up. We trust and like each other, we work well together, we have the right skills and personalities. I think it's important to be open, trusting, and transparent about plans and intentions. We all are aware of the potential upside, and we all appreciate the value of our technology – if successful, it will have significant impact."

Your thoughts on company growth and recruiting new hires?

"At this point they must have the right technical skills and background; important they have passion too, not simply do what is required, need them to constantly think of the company and the team; be a part of our overall vision; in the next twelve months, we plan to add two more people, including a marketing person. We must also aggressively market our technology. Next year will be critical for the company. I see us growing organically with revenues generated from licensing deals; no fast-growth scenario. We plan to develop one family of memory cells that is applicable to many products."

The Board of Directors is made up of Yuniarto, the angel investor, and a third seat that is currently empty. Company A has an observer seat. The angel investor has a semiconductor background – he had started a few technology companies before. There is also a board of advisors with each having an equity share in Zeno "For negotiating agreements, we get plenty of help from Board members and technology advisors. Two of our technology advisors have worked on memory chip licensing before; our law firm also is heavily involved – they provide constant help and feedback on the many legal nuances and subtleties."

Key Relationships

The partnership with Company A is a key relationship – a dedicated partner that provides resources both in terms of investment, equipment, and tacit knowledge to further support the technology development. Also, Zeno recently signed another agreement with Company B, a fabless semiconductor company, for use of its technology in Company B's products.

How many partners can you handle?

"At present, we are looking for strategic partners rather than customers. The question is how many such partners can a small company such as Zeno handle? How many people would we need? How many partners should we have – more partners mean more resources that would require more money – it is somewhat of a dilemma."

Does Stanford play a role in technology or business development?

"We initially built our prototype at the Stanford Nanofabrication Facility (SNF), an NSF-funded center, although almost all the semiconductor processing work is now performed in commercial foundries. We are now developing a 1-megabit memory – it is not possible to use the Stanford facility for this, so we are engaging a commercial foundry. SNF could still be tapped for measurements and testing."

"Stanford Professor Nishi, a technical advisor to Zeno, shares his knowledge as required. When we obtain new silicon data, we meet with him more frequently. He is often too busy, so we find a schedule that works for him with 1-2 weeks' notice. Going forward we will need help for example with soft errors due to radiation effects – Stanford has the facilities to allow us to do this."

What about NSF's role?

"NSF is a crucial factor; I consider it one of the biggest sources of early capital in the semiconductor space; only a handful of VC firms are in this space now. NSF SBIR seed capital for high-risk technology development helps many startups – it is often the only source. If it weren't for NSF, some startups would never get off the ground. The grantee workshops and conferences are very useful for some companies. Dealing with NSF has not been painful at all; progress reporting every six months is good; it helps us too – it's just about right."

Do you participate in trade shows and professional society meetings?

"Through NSF sponsorship, Zeno recently took part in Semicon West – such events help us keep up with the semiconductor technology landscape: we are so absorbed now in developing the technology, we are not ready yet for most trade shows, it's too early but we must remain current."

"We read the published peer-reviewed literature, talk to people to gather insights from their experience. Once or twice a year we attend conferences relevant to memory devices to be aware of technical breakthroughs and for the latest in circuit design, and attend these only if they are held in the U.S."

The Startup Experience

Does control of the company matter?

"Although at this point, I do have majority control, in the long-run it doesn't matter that much to me as long as Zeno is successful."

What are some of the lessons learned, and how would you advise a person starting out?

"Initially I thought it is more about technology, slowly I began to realize it is that and the team and the business development and the market and users – they all complement each other. Maybe I was too naïve – I got to learn about such matters on my own."

"It was difficult in the beginning – I am a quiet person; it wasn't easy to get out of my comfort zone; technical space is my comfort zone. I realized the importance of having a team of different types, each with a different comfort zone. Also, it takes a lot of hard work to build a company and persistence is essential. Something I learned from Board members – there will be ups and downs, but we need to place everything in the right perspective: don't get too down if something bad happens and amplify the good a lot."

You just signed a licensing agreement today – how does it make you feel? What are your key challenges now?

"It was exciting, but there's still so much more work to do. On the technology front, meet all milestones, and on time within budget – if we accomplish them, we can take off quickly; on the business side – we are on the right path with two large agreements in place. It's a good start - we are on the right path with two large agreements in place. It's a good start – we must not mess up; the key is to do the right things at the right time, and especially have the right people; the next 2-3 hires are critical; recruiting people is like closing a deal. We are looking for people all the time."

Have you been lucky?

"I don't know, but luck or prayer – it matters; I would not attribute everything that has happened all to skills and competence. Looking back – I met this guy, this and that happened – many of them are chance encounters. A good example is how I met my angel investor. I was doing consulting work for a small firm and helping with fabricating a device. That person introduced me to this angel. We started talking, he asked me what I do, and one thing led to another, culminating in his investment in Zeno; that day when we met, I didn't go specifically to meet him, didn't even know the person. So, you can call it luck or chance, but it helped get Zeno get started."

Where was this startup at the end of federal government funding?

In 2014, Zeno was able to obtain additional funding for $0.5M from a strategic partner. A license fee payment from another industry partner is due upon meeting specific development milestones. Zeno collaborated with the above strategic partner on the development of Zeno memory technology at 28nm technology node. This partner public-listed company licenses Zeno memory technology, which can provide it a substantial die size reduction and hence cost reduction. This partner will also contribute resources for the collaboration work, which includes design and test resources, as well as fabrication of the 28nm wafers.

The in-kind resources provided enabled Zeno to achieve results without significantly increasing their team size. In addition, the foundry partners were more supportive of Zeno's requests with this strategic partnership than would have been otherwise. It also provides additional credibility and opens path to discussions with other companies and potential licensees.

There are two types of licensing models:

1. License to fabless company – this involves negotiating the terms of the licensing and joint development relationship

2. License to foundry – this involves the design of test-chips in 28nm and 55nm technology nodes in 2016.

Revenues from license fees and non-recurring engineering contracts totaled approximately $4.8M.

Where's the startup now?

Scaling beyond Moore's Law - Zeno Semiconductor, Inc. develops and licenses novel memory and logic technologies, which provide innovative paths to scaling semiconductor devices. The memory and logic technologies, for the company's one-transistor static random-access memory (SRAM) bit cell, are manufacturable on mainstream CMOS and Fin Field Effect Transistor (FinFET) fabrication processes, with no new materials or equipment, and with no changes to any of the existing libraries and IP. Zeno currently has been awarded more than fifty patents on every aspect of its architecture and operation.

Zeno claims that it can boost the drive current of an ON transistor by two times, something that would ordinarily take several process nodes to achieve. And all without increasing leakage, something several process nodes would not likely be able to provide. The smallest SRAM to date fits in the space of a single metal-oxide semiconductor (MOS) transistor, according to serial entrepreneur, Zvi Or-Bach, executive chairman of the newest memory maker in Silicon Valley, Zeno Semiconductor Inc. Zeno Semiconductor unveiled its wares at the International Electron Devices Meeting (IEDM) on December 9th, 2016.

Zeno's 28-nm SRAM bit cells are 37% smaller than Samsung's 10-nm FinFET SRAM bit-cells, which measure 0.04 square microns. They can also be used in either 3-D FinFET or fully-depleted silicon on insulator planar fabs while maintaining their tiny size which scales to even smaller sizes, approaching the sub-nm range, at more advanced nodes. Zeno is also hoping to cover most of the market niches with two models of its novel transistors, one with a single transistor for the ultimate in compactness – five times as many bit-cells as traditional six-transistor SRAMs—and another that adds a second access transistor in series with the memory cell to create a two-transistor memory cell that cuts the leakage current and reduces access time by 40% while still packing three-times as many bit-cells per unit area.

"The two versions of our memory bit-cell enable us to target many different markets. The single transistor Bi-SRAM is targeted at cost-sensitive, low-power applications such as Internet of Things (IoT), while the two transistor Bi-SRAM technology is targeted at high-performance applications such as networking and high-performance computing (HPC)."

Yuniarto claims that the same size advantage can be gained at advance nodes using various architectures, as well as at older nodes for non-demanding applications that are price sensitive. Zeno is also developing the architecture for logic as well as memory.

"We did memory first and we are still concentrating on memory chips, but we are working on logic chips too."

If Zeno can shrink the size of other devices made from its dual-function digital transistors, they may be able to obviate the need to continue seeking more and more advanced nodes to scale devices. Instead of continually attempting to shrink dies ad nauseam, more and more clever combos of MOS with buried intrinsic bipolar functionality might meet many more needs than pure scaling.

Case Study Questions

1. Assess the commercial opportunity and value proposition
2. Business strategy/business model adopted the correct one?
3. Team building – finding enough people with the right skills
4. Identify risk elements and how to mitigate them?
5. What are the areas where Zeno needs help?
6. Develop a growth and exit strategy
7. Start-up valuation exercise – how much is this company worth at this point?
8. Strategic partnering the key? How best to align expectations? How many such partners to have?
9. Discuss the role of NSF in the semiconductor startup space
10. What has been the role of Stanford labs and faculty?
11. Take-away(s)

5 Oceans Provide Sustenance

Technology Sector:	Biotech
Startup:	Stellar Biotechnologies, Inc.
Website:	www.stellarbiotechnologies.com
Location:	Port Hueneme, California
Federal Funding Timeframe:	January 2008 – August 2013
Funding Amount:	$1.20M
This narrative is a story of:	An IPO/reverse merger; strategic partners; bootstrap growth; IP play/issues; significant societal impact; scale-up/production challenges

Company and Team

Stellar Biotechnologies, Inc. ("Stellar"), a publicly held biological manufacturing corporation, was formed in 1999 to develop technology and manufacturing capabilities for production of marine natural products for the pharmaceutical industry. The company founders were pioneers in the development and commercialization of abalone aquaculture in California. They recognized an opportunity to apply their successful experience in aquaculture to the development of commercially sustainable supplies of important marine-derived biomedical products. Stellar is led by experienced senior management and research scientists from the marine aquaculture and biomedical industries, supported by an operations team with skills in pharmaceutical manufacturing, quality assurance, and commercial aquaculture production.

Stellar specializes in the production of a critical component of a new class of medicines known as therapeutic vaccines. The company's products are formulations of Keyhole Limpet Hemocyanin (KLH), a previously resource-limited highly immunogenic carrier protein used in the production of vaccines for cancer, hypertension, arthritis, and other serious diseases. This protein is an important immune-stimulating molecule widely used as an active pharmaceutical ingredient in many new immunotherapies, and as an injectable product to assess immune response. This versatile molecule has a long history of safe and effective use across a wide range of disease indications and research applications. A paradigm shift in drug research toward treatment strategies that focus the body's own immune system to target disease led to rising demand for Stellar's KLH platform.

Stellar applied decades of specialized aquaculture science to a pharmaceutical industry challenge and created the only KLH production facility of its kind in the world. The company's customers and partners include multinational pharmaceutical companies, renowned research centers, and developers of active immunotherapies and therapeutic vaccines. The company transformed a threatened natural resource into a viable commercial platform - a portfolio of products including carrier protein for vaccine conjugation, finished product for immune stimulation in immuno-toxicology applications, and novel assays (an investigative (analytic) procedure in laboratory medicine, pharmacology, environmental biology and molecular biology for qualitatively assessing or quantitatively measuring the presence, amount, or functional activity of a target entity (the analyte)). The company's competitive advantages are its sustainable and proprietary methods for commercially scalable production of the fully purified, regulatory approved, Good Manufacturing Practices-grade KLH products.

The company's vision includes production of additional marine species for drug discovery and development, development of cell culture, fermentation, or synthetic pathways for production of marine-derived biologics, and ultimately the manufacture of active pharmaceutical ingredients and finished drug substances using its KLH products, contract manufacturing relationships, and distribution alliances. The method for non-lethal hemolymph extraction was developed and patented soon after identifying initial products and customers. This extraction method allows many years of repeated commercial KLH harvest from the same animals, offering a sustainable, highly consistent, and scalable supply of KLH to meet the growing pharmaceutical market demand. Additional IP includes proven husbandry methods for adult keyhole limpets under controlled aquaculture conditions that promote KLH synthesis for commercial harvest and sexual maturation for hatchery production of successive generations of production animals.

The Stellar team has deep experience in developing and financing biotech and pharmaceutical companies, and previously held senior positions at successful public companies such as Genentech, Johnson & Johnson, and Abbott Labs, and academic institutions like Harvard, Scripps Research, and MIT. The company currently has sixteen full time employees. Frank Oakes, Chairman and CEO, has more than thirty years of management experience in aquaculture including a decade as CEO of The Abalone Farm, Inc., during which he led that company through the R&D, capitalization, and commercialization phases of development to become the first profitable and largest abalone producer in America. He is the inventor of Stellar's patented method for keyhole limpet hemolymph extraction.

Dr. Catherine Brisson, Chief Operating Officer, has more than twenty years of experience in the biotechnology, pharmaceutical and medical device industries with broad scientific and operational expertise in the areas of quality assurance, quality control, regulatory affairs, manufacturing, and product development. Kathi Niffenegger, Chief Financial Officer, has more than thirty years of experience in accounting and finance. Kathi was previously technical partner in the audit division of Glenn Burdette. Mark McPartland, VP of Corporate Development and Communications, has more than sixteen years of experience in business development, capital markets advisory, corporate communications, and C-suite consulting. Prior to joining Stellar, he served as Senior Vice President at MZ Group, a subsidiary of the world's largest independent global investor relations consulting firm.

Market Opportunity

KLH is a protein derived from the blood of the keyhole limpet, a marine mollusk, a group of invertebrates that includes squids, octopuses, snails, and mussels. This protein is potently immunogenic yet safe in humans and therefore highly valued as a vaccine carrier protein. It cannot be synthesized and is available only as a purified biological product from the mollusk. KLH is a critical component of several advanced therapeutic vaccines for lymphoma, bladder, breast, colon, and other cancers. The rapidly growing interest in therapeutic vaccines and its documented efficacy are creating a significant biopharmaceutical market for this molecule. Over a dozen companies are currently testing KLH-based cancer vaccines in human clinical trials. For successful commercialization, vaccine developers will require reliable and commercially scalable supplies of this critical component. As vaccines obtain regulatory approval, there will be increasing need for this protein made according to the Food and Drug Administration (FDA) Good Manufacturing Practices (GMP).

In today's supply environment, a single successful vaccine product could result in a shortage of GMP-grade KLH, a product supplied principally by Stellar and its main competitors, Biosyn and Sigma Aldrich. These competitors rely on limpets found only sporadically in the coastal region from central to northern Baja California; the animals are typically bled to death to obtain the KLH-containing hemolymph. The commercial prospects of KLH vaccines are thus threatened by the industry's reliance upon a limited, naturally fragile and unsustainable population of source animals. With the imminent potential approval of KLH-based vaccines and increasing KLH demand, Stellar's solution to this supply problem is the aquaculture production of the keyhole limpet. The company has manufactured and produced commercial lots and initiated long-term stability studies. Unlike its competitors, Stellar has a unique, patent-protected ability to manufacture consistent, sustainable supplies of KLH from the only source of the molecule, the keyhole limpet, from a controlled aquaculture environment.

The key driver of the company's business is the emerging market for KLH-based therapeutic cancer vaccines. Oncology is one of the fastest growing segments of the pharmaceutical industry, due in large part to the outstanding clinical and commercial successes of recent antibody therapies for cancer, which together generate revenues of $10B per year. The demographics of aging populations and the proliferation of target molecule discoveries for new drugs resulting from advances in genomics and proteomics are generating commercial interest and investment in new cancer immunotherapies, including vaccines of promising tumor antigens conjugated to KLH. The enthusiasm for KLH-based cancer vaccines is also the result of favorable clinical trial results – they are proving to be valuable in prolonging remissions. The treatment of tumor recurrence following the failure of conventional therapies such as chemotherapy, radiation, and surgery remains a significant clinical problem, and a promising area for therapeutic vaccines.

In the therapeutic vaccine market, Stellar is competing with an established competitor with a product in clinical trials. Displacing this competitor will require selling customers on the compelling value proposition of the firm's sustainable KLH supplies and demonstrating the bio-equivalence of its KLH to the competing product. In the cancer vaccine market, the company has established strategic relationships and engaged potential customers without the use of distributors or advertising. The firm negotiated a co-marketing relationship with a leading contract manufacturing organization (CMO), an established provider of GMP conjugation services, to act as a conduit for Stellar's product to new biopharmaceutical customers.

Currently, Stellar markets two GMP-grade KLH products to the biopharmaceutical and vaccine development markets. It has supply agreements with multiple vaccine developers. In the near-term the focus will be on the growing KLH-based therapeutic vaccine market. They will serve this market with a mix of finished (sterile vialed) KLH products for use by biopharmaceutical companies, as well as their traditional bulk KLH intermediates that are sold to contract pharmaceutical manufacturers to produce finished vaccine products. Scalability and insulation from resource limitations are acknowledged as the company's longer-term advantages that address the critical constraints that might emerge during scale-up for vaccine commercialization. The company is also pursuing customers in the preclinical stages of vaccine development that do not require bioequivalence testing, and therefore have lower barriers to product acceptance.

The strategic advantages offered by Stellar are sustainability, control of supply, and consistent high quality, market advantages that are readily acknowledged. The company's existing and potential customers have identified the improved lot-to-lot consistency and reduced regulatory risk of KLH from a controlled aquaculture source versus uncontrolled ocean sources potentially subject to environmental contamination as key factors in choosing Stellar as a KLH supplier. This will also help to insulate Stellar and its customers from the risks associated with environmental change and resource declines.

A niche point-of-entry for new vaccine development are researchers from academic, medical, military, and veterinary institutions that do not have ready access to proprietary vaccine technologies. For them, KLH is a familiar and convenient platform for testing new vaccine candidates. Stellar's product offers the added benefit of suitability for clinical testing in humans. The availability of a commercial GMP-grade KLH source will also facilitate out-licensing and commercial development of new vaccines developed by the company's institutional customers. By 2014, because of marketing approval of KLH-based cancer vaccines, the total KLH vaccine market will be $3.5B requiring several kilograms of KLH. Among the various KLH-based therapeutic vaccines under development, several are potential "blockbusters", i.e., capable of generating $1B in annual revenue. FDA approval of the first KLH vaccine will stimulate the market by validating the use of KLH for vaccines and thus encourage further development of new KLH-based therapeutics.

Technology and Product

Technology and product development efforts are focused on developing and optimizing effective hatchery methods for settlement and metamorphosis of keyhole limpet larvae. Unlike other cultivated herbivorous molluscan species, keyhole limpets are opportunistic carnivores and predisposed to cannibalism. Knowledge of the underlying biochemical factors that promote settlement, metamorphosis and early post-larval survival is critical for controlling production[11]. Current efforts involve translating research results into designs for testing and optimization of systems, diets, and aquaculture methods for cultivation of the age-specific developmental phases, from metamorphosis to fully developed juveniles capable of being transitioned onto adult diets and cultured for KLH production.

The goal is to develop reliable methods for producing keyhole limpet at a scale sufficient to support the production of commercial quantities of KLH, and entirely from animals spawned and reared in a controlled-environment aquaculture system. It is to demonstrate in animal model systems the basis for functional differences observed between different KLH formulations, characterize their molecular and cellular immune mechanisms as required by customers and strategic partners, and meet anticipated U.S. FDA requirements. Additional technology roadmap elements include product and process development, diagnostic assay development, manufacturing, analytical methods development, and pre-clinical safety testing protocols to support FDA filings for KLH-based diagnostics.

Stellar has developed the world's only dedicated aquaculture facility, and non-lethal extraction and manufacturing technologies for sustainable GMP-grade KLH. Production is based on a modular, scalable system which allows capacity to be incrementally added as KLH demand increases. It involves keyhole limpet acquisition or hatchery production, husbandry, hemolymph extraction and purification to produce bulk intermediates, GMP-grade KLH purification to produce bulk finished products, and fill-finish to produce sterile vials of final products. Stellar has developed proprietary methods for each of these manufacturing steps. Manufacturing costs for the KLH intermediate is based on production history. Finished product GMP-grade KLH manufacturing, testing and fill-finish are currently performed by contract manufacturing organizations.

The company's KLH must also demonstrate bioequivalence - biological efficacy such as can be shown by *in vivo* comparability studies of immune response. This is addressed by extensively characterizing physico-chemical properties, its conjugation potential, and its immunological potency, and comparing these product attributes with that of competing products. Bayer Innovation GmbH (BIG), a subsidiary of Bayer Healthcare AG, selected KLH as the immunogenic protein carrier for its vaccine technology and thus has a strategic interest in securing a long-term supply from a safe and sustainable source. Based on the strength of Stellar's supply proposition and the support for technology development provided by NSF SBIR funding, Bayer entered into a joint development agreement (JDA) with the company in 2009. That agreement specifically sought to optimize methods for purification of KLH to address issues of commercial scale manufacturing and to increase safety and efficacy.

There is a need for tests to determine the ability of immune compromised patients to respond to a new protein antigen such as KLH. Immune status testing is relevant in determining whether these patients are capable of effectively responding to vaccines as well as infectious agents such as bacterial and viral pathogens. Standardization of KLH testing with well-characterized antigen and test kits is desirable. Studies have been conducted to immunologically characterize different KLH forms for use in such tests. To develop its proprietary KLH formulation for use in immune function testing, the company is conducting additional animal model studies to unequivocally establish various preferred formulations. Such studies are critical for defining the functional properties of the different KLH formulations to critically establish the correct product for different clinical uses.

Intellectual property (IP) in this space includes method and composition-of-matter patents, and trade secret combinations of public domain methods for KLH purification and formulation. Products are differentiated based on specifications, GMP compliance, and by source -

aquaculture vs. natural resource. First-to-market producers are at an advantage because new suppliers must meet or exceed the specifications established by existing suppliers and show bioequivalence for their products. The firm's IP includes proprietary aquaculture methods and culture systems for all life-cycle stages of the source animal, a patented non-lethal hemolymph extraction process, and proprietary protein purification methods. Its aquaculture methods and KLH purification and testing methods are trade secrets. As Stellar enters the retail market with its clinical reagents and combination products, brand identity will become increasingly important. The company anticipates establishing trademark protection for future branded products.

Business Model and Execution

The business strategy is to sell to vaccine developers as early as possible during the development cycle, under long-term supply agreements that offer assured scalability and continuity of supply. Customers include commercial cancer vaccine developers and clinical researchers with a need for KLH that effectively addresses the safety concerns and quality standards for products of biologic origin required by drug regulatory agencies in the U.S. and Europe. The company is currently focused on developing customers for its products among the companies that have already invested in KLH-based vaccine development, especially those with vaccines in clinical trials.

Stellar markets KLH products directly to biopharmaceutical customers. Company managers have formed relationships with major players in the cancer vaccine market. The company already has sales of KLH intermediates and purified formulations derived from their captive colony of ocean-harvested limpets priced to be competitive with existing suppliers. Increasing demand will drive KLH prices upward, and this trend will become more pronounced as natural (non-aquaculture) supplies become scarce. This is likely to increase the value proposition of the firm's sustainable KLH products over those competitors who do not have the advantages of aquaculture technology. The growth plan is built upon its demonstrated ability to win customers for its KLH products by promoting the sustainable supply advantages offered by the company's unique aquaculture technology for cultivation of the source animal.

The strategic advantages readily acknowledged in the marketplace for the company's aquaculture-derived, GMP-grade KLH are sustainability, control of supply, and consistent high quality. Its products are further distinguished by the fact that they are produced by repeated, periodic, non-lethal extraction from captive colonies grown in a controlled, GMP-compliant aquaculture system. To derive further value from its aquaculture based KLH product line, Stellar also plans to develop GMP conjugate manufacturing services for its KLH carrier protein products. This service will leverage its growing expertise in biopharmaceutical manufacturing, and in the preparation of peptide and carbohydrates conjugates for clinical trials. This long-term development plan may require the company to invest in its own GMP manufacturing facility.

Competitors in the KLH supply business are specialty pharmaceutical manufacturers marketing a proprietary vaccine platform; and biotech/pharmaceutical companies with interests in vaccine platform technologies. A vital component of the firm's competitive strategy is developing KLH deemed by regulatory authorities to be bioequivalent to that already being studied in clinical trials. The ability to show bioequivalence is a critical success factor for the company's ability to displace competing KLH products and to qualify as a source of supply for existing vaccine products. Stellar is also pursuing customers that are in the early (preclinical) stages of vaccine development which do not require bioequivalence testing, and therefore have lower barriers to product acceptance.

The total NSF SBIR funding to Stellar in the period 2008-2013 has been approximately $1.1M. The company's value proposition, substantiated by the NSF SBIR funding for aquaculture technology development and the joint development agreement with a strategic pharmaceutical partner supported a successful 2010 financing of approximately $8M. The funding was specifically structured to support aquaculture KLH commercialization by offering common shares for public trading on the TSX venture exchange in Vancouver, Canada. Liquidity for company shareholders will most likely be through an acquisition by a larger player in this market segment.

NSF Commercialization Assistance and Impact

Stellar was introduced to IA in 2009 by their NSF Program Director. The company's needs as identified by IA converged upon the company's commercialization efforts. Contemplated issues included help with bankers, and introductions to financial entities. The company started to seriously engage with IA in the summer of 2009 as they were exploring the opportunity to go public on the Toronto Stock Exchange via a reverse IPO. IA introduced Frank Oakes, the company's CEO, to Mike Zarriello, Executive Director, Capstone Advisory Group, to act as an advisor during this process. The value of Mike's advice throughout this process was recognized by Frank in the form of an invitation to sit on the company's Board of Directors; Zarriello declined. Additionally, IA introduced John Sundsmo as an "at-large" advisor to the company in early 2010. John later took on the role of Vice President of Research and IP Management. IA continues to monitor the company's growth and responds to requests from the company as needed.

In January 2011, the inaugural "IA @" event series was held in conjunction with the "Nuts & Bolts" entrepreneurial course at MIT. Three NSF SBIR-funded startups were selected to present as innovation 'case studies', Stellar being one of them. Most of the MIT students who attended and participated in the question-and-answer sessions were not aware of the support and opportunities provided by the NSF SBIR program. As a direct result of the event, at least two MIT startups submitted SBIR proposals. One of them, Arctic Sand, a later NSF SBIR grant recipient, went on to secure $9.6M in Series A financing at the end of 2012 – investors included Arsenal Venture Partners, Ray Stata, an angel investor, and other strategic investors. Additionally, IA gathered participants from venture, corporate, and policy networks to foster knowledge of NSF programs and get exposed to the many technologies being developed by startups funded by the NSF SBIR program.

IA activities and outcomes are summarized in the following Table.

IA Activity	Outcomes
Startup Overall Engagement	49 months; 150+ hours of interaction; 150+ e-mails sent to and on behalf of Stellar
Recruit Management Team	John Sundsmo, Ph.D., named to VP, Research & IP Management
Operational Assistance	Stellar wished to avoid the dilution typical in the biotech space by going public on a foreign exchange through a reverse IPO; it lacked experience doing so; IA introduced Stellar to Mike Zarriello, an experience financier, who has executed multiple reverse IPOs and was previously the lead banker when Nike bought Cole Hahn; Stellar has benefited from Zarriello's advice – he also provided financing introductions and guidance
Find Investors	Capstone Advisory Group
Identify Domain Experts	Michael Zarriello; John Sundsmo

Where was this startup at the end of federal government funding?

In 2012, Stellar raised $3.1M through a reverse merger with a Canadian Capital Pool Company (CPC). In conjunction with its NSF Phase IIB research, Stellar used the funds provided by its partner Bayer Innovation GmbH (BIG), to advance the development of manufacturing methods for GMP-grade suKLH, the product formulation most requested by vaccine customers. BIG has a commercial need for KLH and a strategic interest in securing a long-term KLH supply from a safe and sustainable source. The optimization of the manufacturing methods for the suKLH formulation further allowed the company to establish a market for KLH produced from aquaculture M. crenulata as a carrier protein for KLH conjugate vaccines. In addition, the company is now using funds from its Series A & B financing enabled by the NSF matching supplement to develop a HMW KLH formulation for use as an immunogen and neoantigen for immune response testing. Funds from these financings were used for KLH product development, assay development and manufacturing. Through these efforts the Company is now offering a suite of products including multiple formulations of KLH for clinical and research use and a suite of KLH diagnostic kits.

The company is currently actively marketing Stellar KLH™ branded products for pharmaceutical and research use produced from aquaculture technology derived from this NSF-funded research. The product list includes:

1. Keyhole Limpet Hemocyanin (KLH20MV), GMP-grade subunit, KLHKLH for protein conjugation and carrier protein applications

2. Keyhole Limpet Hemocyanin (KLH01NV), GMP-grade high molecular weight KLH for immune function testing applications

3. Keyhole Limpet Hemocyanin (KLH20MVR), high purity research--grade subunit KLH for use as a multivalent adjuvant conjugate

4. Anti-KLH enzyme-linked immunosorbent assay (ELISA) for pharmaceutical immunotoxicology research

The number of employees rose from six to fifteen over the course of this federal funding, and revenues totaled approximately $200K. Customers for the company's products are vaccine developers, clinical researchers, biotechnology and pharmaceutical companies, and contract research organizations (CRO). The marketing approach has been to generate product and brand awareness through direct contact with companies currently using KLH in vaccine development and immune function testing, presentations at scientific conferences, e-mail distribution of product information, and through the development of a KLH information website, KLHsite.com. Because sales in this industry are driven by scientific data rather than product promotion, the company's goal has been to generate and present scientific data supporting the product value proposition. In many cases this effort has led to collaborative research with potential customers that require specific research data support the integration of Stellar KLH™ into vaccines in development or immune function testing protocols.

The commercial products produced directly or indirectly from this NSF SBIR research consist of purified KLH in various formulations and diagnostic test kits (Anti-KLH ELISA) that use KLH as an essential manufacturing reagent. The market for Stellar's products is composed of three distinct segments:

- The therapeutic vaccine segment of the biotechnology and pharmaceutical industries. Potential customers: vaccine developers using KLH as a carrier protein and/or adjuvant.

- The immune response assay segment of the medical diagnostic industry. Potential customers: CROs, pharmaceutical and biotechnology companies performing T-cell dependent antibody response (TDAR) testing.

- Research institutions and companies performing preclinical drug development and immune function research. Potential customers: university laboratories, biotech, and medical research institutions.

Where's the startup now?

Based north of Los Angeles at the Port of Hueneme, Stellar Biotechnologies, Inc. (Nasdaq: SBOT) is the leader in sustainable manufacture of Keyhole Limpet Hemocyanin (KLH), an important immune-stimulating protein used in wide-ranging therapeutic and diagnostic markets. KLH is both an active pharmaceutical ingredient (API) in many new immunotherapies (targeting cancer, immune disorders, Alzheimer's, and inflammatory diseases) as well as a finished product for measuring immune status. Stellar is unique in its proprietary methods, facilities, and KLH technology. The company is committed to meeting the growing demand for commercial-scale supplies of GMP grade KLH, ensuring environmentally sound KLH production, and developing KLH-based active immunotherapies. Stellar KLH is a trademark of Stellar Biotechnologies.

During the first nine months of the year, the company reported continuing progress across its initiatives to optimize and expand its manufacturing capacity at its primary production facility in California. Stellar's current manufacturing systems, which were originally developed to provide clinical-stage quantities of its KLH carrier molecule, currently support multiple KLH therapies in advanced clinical studies. These studies include Alzheimer's disease, lupus and metastatic breast cancer.

"We now have multiple customers with Phase II clinical studies that are fully enrolled and underway, and we are preparing for the impact that favorable clinical results could have on the KLH market and our supply capabilities," said Stellar President and CEO Frank Oakes. "We plan to incrementally increase our production capacity and manufacturing capabilities in line with the needs of our customers and to maintain our leadership position in the sustainable production of controlled, fully traceable KLH." Stellar Chief Financial Officer Kathi Niffenegger noted that research and development activities increased under the company's optimization plan for the three and nine months ended June 30, 2017. "In the third quarter, we continued the trend of prudently managing our working capital. We reduced corporate expenses and redirected resources to continue the ramp-up of our development activities. In addition to the operational benefits of increased throughput capacity, we believe these optimizations will favorably impact our financial metrics at the higher production volumes needed for commercial drug launches."

Total revenues were $0.02 million for the quarter ended June 30, 2017, compared to $0.18 million for same period last year. The change was due to a decrease in product sales volume. While the company's customer base has not changed significantly, product sales volumes are subject to variability associated with the rate of development and progression of clinical studies of third-party products that utilize Stellar KLH. On June 30, 2017, the company had working capital of approximately $7.7M. Cash, cash equivalents and short-term investments totaled $7.65 million.

Stellar will file its Form 10-Q for the quarter ended June 30, 2017, with the Securities and Exchange Commission (SEC) on August 9, 2017.

Case Study Questions

1. Assess commercial opportunity/potential and value proposition
2. Comment on technology strategy/roadmap
3. What are the barriers to entry for competitors?
4. Discuss the company's growth strategy – e.g., sales-driven steady and slow growth, M&A, spin-off, IPO
5. Potential risk elements
6. Regulatory issues
7. Strategic partner relations – communications/clarity; expectations alignment
8. Take-away(s)

6   No Need for Too Much Lubricant

Technology Sector: Manufacturing
Startup: Fusion Coolant Systems, Inc.
Website: www.fusioncoolant.com
Location: Ypsilanti, Michigan

Federal Funding Timeframe: January 2010 – June 2014
Funding Amount: $0.88M

This narrative is a story of: Bootstrap growth; strategic partners; scale-up/production challenges; location and technology sector, a hindrance to raising equity capital

Company and Team

The $2.7 trillion U.S. manufacturing sector requires higher productivity and profitability to remain competitive. Fusion Coolant Systems ("Fusion") is commercializing technology which enables long-term cost savings and environmental benefits in the manufacture of machined parts. Manufacturers have used various forms of metalworking fluids (MWF) to improve the performance of metal working processes such as machining, grinding, and forming. Oil- or water-based coolants or oil/water emulsions are necessary for removal of extreme heat generated during such operations. Many systems in use today are expensive, ineffective, and carry substantial risk for employee health and safety. Coolant/lubricant systems are a critical part of manufacturing processes, but beyond incremental changes to existing systems, there has been little technical advancement in the area for several decades.

Fusion, a Michigan-registered corporation, was spun out of the University of Michigan (UM) in 2009 by Steven Skerlos and Andrew McColm. The startup's core intellectual property (IP) was developed by Steven through research funded by the National Science Foundation (NSF) and the Environmental Protection Agency (EPA). The company's mission is to maximize the performance of machining processes using its Composite High-Pressure Lubrication (CHiP Lube) technology based on supercritical carbon dioxide. This technology significantly reduces operational costs, effluent streams, and potential health risks for end-users. The U.S. Midwest has a long and deep supply of firms with expertise in custom and semi-custom design, development, and fabrication. Manufacture of the CHiP Lube system is subcontracted to a Michigan-based custom hydraulics engineering and fabrication firm that was chosen through a competitive bid process. Discussions are underway with firms that provide toll-blending services for conventional metalworking fluids to secure a long-term supplier of the proprietary lubricants required by this lubricating system. Fusion will retain several core activities and competencies including design and development of the lubricant, sub-systems and overall system, and specialized tooling.

Thomas Gross, CEO, was the former General Manager of Gleason Works, and the former Chief Operating Officer (COO) of Cross and Trecker. Steven Skerlos, Chief Technology Officer (CTO) and scientific advisor, is a distinguished Professor of Mechanical and Environmental Engineering at UM and inventor of the technology. His prior startup, Accuri Flow Cytometers, received four rounds of venture capital funding and is today a thriving, growing business. Andrew McColm was the interim CEO and Chief Financial Officer (CFO) - he has been a founder/senior manager at half-a-dozen startup companies across three continents. During his tenure as Associate Director, UM Office of Technology Transfer, Andrew was involved in the formation of over three dozen technology-based companies across a range of industries from software and materials to medical devices. Scott Jones, Engineering Director, has an extensive background in product development through various positions at General Motors (GM). He was involved in the creation and management of three high-tech startups as Adam's Entrepreneurial Fellow at Wayne State University. Other members of the team are David Stephenson, Research Scientist, an industry veteran in machining and machine tools. The author of several books on machining, he was a senior research engineer at GM. James Giovanni has over two decades of sales and marketing experience in the machine tools and automotive industries. Andrew King, a senior design engineer, developed the initial testing version of the CHiP Lube coolant control system. He works part-time with Fusion through his prototyping firm, Kingtech.

Terry Cross, Board Member, is a serial entrepreneur and active angel investor. Terry entered the machine tool business in Detroit with The Cross Company, the world's leader in automated machine tools. He then spent the next twenty-five years in the brokerage and investment banking business. Terry has invested in forty private companies including Novell, PayPal, Napster, and was a first-round investor in Google. Legal counsel is provided by David Parsegian of Honigman Miller who has extensive experience in startup company formation, private equity, and venture capital.

Market Opportunity

Metalworking Fluids (MWF) are designed to cool and lubricate manufacturing operations. This $1B industry is highly fragmented with the top ten firms holding a cumulative market share of 40%. The remaining

60% is split between hundreds of smaller firms. The top-tier suppliers include Castrol Industrial, Houghton International, and Quaker Chemical. The U.S. MWF market is challenged by limited technological advancements and the increasing use of non-ferrous alloys such as aluminum and titanium, especially in the aerospace industry. Introduction of high-speed cutting and grinding tools in many end applications have led to problems with foaming, misting, and mediocre performance with traditional MWFs. Suppliers have been scrambling to modify their fluid chemistries or adopt new strategies to cool and lubricate rolling and machining operations to address this problem. Additionally, rising base oil, raw material, energy, and transportation costs reduce profit margins. Stringent U.S. health, safety, and environmental regulations have forced suppliers to develop environmentally compliant products that meet customer requirements.

Current MWF products do not meet the increasingly rigorous demands of machining modern materials such as aerospace alloys, compacted graphite iron (CGI), and high silicon steel and aluminum alloys. These products, emulsions of oil and water, represent a continuous tradeoff between cooling (more water) and lubricity (more oil). MWFs are highly susceptible to contamination by microorganisms leading to potential health risks for workers. Not being environmentally benign, their disposal is expensive. The Fusion CHiP lube system is the next evolution in MWFs that addresses these new market realities. It provides a new paradigm in metalworking fluids allowing both cooling and lubricity to be simultaneously increased. The increased metal removal rates and tool life also lead to substantial reductions in operational costs. It uses non-toxic and bacteria-free components, therefore reducing health and safety risks for workers, and reduces the disposal of MWFs minimizing potential environmental hazards.

Customers that manufacture either raw material or finished parts that require MWFs in various volumes can be classified into three main groups. The first is industrial customers such as the large metal forming mills, automotive manufacturers, suppliers, and capital equipment manufacturers. These large customers have a sophisticated approach to manufacturing technology and are large purchasers of MWFs of more than 100,000 gallons per year. This group's rolling and machine tool needs are normally serviced directly by dedicated representatives from machine tool manufacturers. The second group consists of manufacturers with fewer machines whose principal function is metalworking such as specialty aerospace manufacturers that work with advanced alloys in higher-margin, lower-volume operations. They are sophisticated about machining functions but have less in-house infrastructure and expertise in handling innovative technologies and innovations. This group is typically served through a dealer network by machine tool manufacturers. The third group consists of small job shops with ten machines or less. The cost to service this group of customers is high in proportion to potential revenue accrual. It would therefore be of lower priority for Fusion unless specific high-value applications can be identified.

Market barriers are slow adoption and long lead-time for new equipment purchases, and step-function technological changes broad adoptions of which take longer. The firm's objective is to have twenty customers belonging to the first group and forty customers in the two secondary groups within five years from the initiation of sales. This rollout anticipates the lengthy lead-times required to take innovations from lab to production floor in large metal forming and manufacturing companies. Per customer revenue from tier-1 customers is anticipated to eventually grow to be several million dollars per customer per year. Fusion will initially target high-value opportunities - metal forming applications where significant throughput and cost-saving benefits provide immediate return-on-investment (ROI), where current solutions underperform, and where the organization is looking to mitigate environmental and health concerns with solutions that add to the bottom line.

Industrial partners have identified specific metal-forming and machining operations that cause throughput limitations, unacceptable MWF treatment and recycling costs, and excessive tool wear using today's MWFs. A large producer of aluminum that meets the above criteria has been identified as a target customer. Discussions with company representatives at various levels of management and engineering have yielded significant interest in Fusion's technology. This potential customer would like to retrofit their rolling mills that currently use mineral oil coolants with CHiP Lube. This company believes this can lead to substantial cost-savings in operating expenses and positively impact their carbon footprint. Other applications which fit these criteria are commonly found in mid-size aerospace contractors and other established manufacturing installations that are developing new processes using advanced alloys. One potential customer, a Michigan-based company, manufactures intricate, high-precision aircraft engine components. In improving manufacturability of these high-value advanced alloys, CHiP Lube can provide a large benefit in proportion to the system cost thereby reducing barriers to adoption.

Technology and Product

The patented CHiP Lube technology is a radically different solution from conventional water-based systems for machine tool cooling and lubricity. It is the most significant advancement in metal working fluids in several decades and has the potential to address many parts of the value-chain: coolant delivery equipment, lubricants, services, and specialized tooling and tool holders. CHiP Lube is composed of naturally occurring, renewable components and is therefore environmentally sustainable. It enables Minimum Quantity Lubrication (MQL) by reducing coolant usage and allowing for a single versus multiple pass system so recycling and reconditioning of MWFs is eliminated[12]. This represents a significant cost-savings for the end-user. System components are also non-toxic and bacteria-free thus improving worker health and safety. Finally, this system improves both lubricity and cooling thereby allowing for dramatic increases in metal removal rates and tool life leading to substantial reductions in overall manufacturing cost.

CHiP Lube technology consists of two components, supercritical carbon dioxide and a vegetable oil-based lubricant, combined to form a single-phase cutting fluid. Lubricant solubility in supercritical carbon dioxide is a key feature that enables the fluid to provide unparalleled cooling and lubrication simultaneously. This allows increases in metal removal rates of over 50% and/or doubling tool life depending on the application. NSF Small Business Innovation Research (SBIR) funding allowed the company to develop and build a production-level delivery and control system for field trials, develop a system for variable lubricant coolant control, and integrate the system for on-site pilot testing at a customer site in partnership with the NSF-funded University of Michigan Engineering Research Center (ERC). For practical implementation of this supercritical carbon dioxide based MWF, university researchers and then Fusion engineers designed systems for the control and delivery of the composite CHiP Lube metalworking fluid. The system can then be used externally for rolling, extruding, forming, turning, grinding, and sawing applications, or through-tool for drilling and milling.

Initial capital investment and operating cost models with various customers show significant cost-savings compared to traditional MWF systems with short payback periods and high ROI. One customer has estimated annual savings of $3M per rolling mill per year by replacing their current mineral oil technology with the CHiP Lube system. Another customer has estimated that they could save eight dollars per part just on tool cost in a large-volume assembly line by switching to CHiP Lube. The part is made of a material that is notoriously hard to machine with severe tool wear under existing machining conditions. Initial tests have shown that the superior lubricity and cooling provides substantially reduced tool wear. The savings here are not from the elimination of the conventional fluid coolants but rather in reductions in tool wear and in increased throughput. The common impediment to product rollout in both the above cases is that neither of them is willing to risk full-scale implementation with technology that is not fully proven. Large, sophisticated customers have long procurement cycles and require considerable product validation yet are open to innovative manufacturing methods. They generally have in-house expertise to perform detailed technical and economic evaluations of innovative technologies. They also understand the total cost of ownership of their existing MWF solution thereby making the economic justification to use CHiP Lube easier. They are thus ideal initial customers in terms of their willingness to adopt but some of the worst in terms of their speed of adoption.

Due to its strong solvent qualities, carbon dioxide and supercritical carbon dioxide have been used for decades in applications such as cleaning and chemical extraction. However, the application of supercritical carbon dioxide to machining as developed by Dr. Skerlos and his colleagues is unique and is protected by a 2008-issued U.S. patent and nationalized in several countries in Europe, and in Japan and Canada. An exclusive license agreement for all fields of use and all geographic areas has been negotiated with the University of Michigan. A U.S.-based company has attempted to patent the composition of a low-pressure carbon dioxide-based fluid for machining applications, but it was rejected on all claims based on prior art. Several patents exist for low-pressure applications of carbon dioxide for machining applications, but none relate to the use of high-pressure supercritical carbon dioxide. The key advantage of supercritical carbon dioxide is its ability to dissolve a high lubricant load at a much higher mass transfer rate thereby leading to both superior cooling and lubricity performance. It is anticipated that follow-on IP will be in delivery components and/or in application specific tools and accessories.

Business Model and Execution

Startup funds consisted of capital contributions from the two founders, state and federal grants, and employees accepting below market salaries in exchange for equity. Total NSF SBIR funding to Fusion Coolant Systems in the period 2010-2014 amounts to approximately $875K. Given the nature of the technology and the firm's location in the Midwest, there are potential angel investors who understand the benefits to lowering manufacturing costs and have the wherewithal to make a substantial investment. Although senior management has had numerous discussions with venture capital firms, it is felt that a startup in the manufacturing space typically finds it hard to attract such investment. In 2013, the company raised $600K in financing from The Frankel Fund, a University of Michigan student-led venture capital fund, as the lead investor. Once the company's market capitalization reaches $30-50M, it could be an attractive acquisition for a larger machine tool company or mid-market private equity firm.

Company growth will be driven by customer revenue. Management's objective is to build-out a long-term sustainable business that generates initial revenue through equipment sales/leases and ongoing revenues through long-term contracts for maintenance and consumables. Management's ability to secure proof-of-concept development contracts and subsequent sales will drive initial revenues. The company's manufacturing strategy is to outsource all manufacturing operations to specialty firms thereby minimizing initial capital investment and preserving flexibility of design. Commensurate with the company's growth will be a corresponding reduction in the machining costs of their U.S. customer base. As CHiP Lube becomes established in the metal-forming and machining areas, the product line will be expanded to include value-added services and specialty tooling. International expansion through partner machine tool companies and international operations of domestic customers is also anticipated. Financial projections are based on actual customer requirements developed through extensive market research and several customer interviews. The first market segment with significant sales will be large- and medium-size customers due to anticipated quick payback periods for investing in the technology and the extremely high ROI for such companies.

Market opportunities broadly consist of retrofitting the technology onto machines in place of existing water or MQL-based systems, and the integration of the CHiP Lube system directly into new machine tool purchases. Revenue generation from these two opportunities will consist of the following: the cost of retrofits; equipment sale or lease; one-time installation and training fees; ongoing site license and material supply for lubricants and other consumables such as specialized tools and tool holders; and original equipment manufacturer (OEM) system integration. Sublicensing revenue from machine tool manufactures can also be a revenue source. Relationships are being developed with several machine tool manufacturers to pre-equip machines with the CHiP Lube option. Currently, the company is pursuing sales of its first-generation CHiP Lube delivery hardware to target customers for development and evaluation purposes.

Long-term leasing of capital equipment will also be offered, probably through a third-party financial partner. This option will be attractive to market segments currently contracting out metalworking fluid management to outside firms. Supply of proprietary lubricants and ongoing maintenance of equipment will be long-term revenue sources. Due to the very low viscosity and high solvent quality of the supercritical carbon dioxide, high pressure pumps for the system will need to be serviced at intervals of two to four months. Also, use of non-Fusion lubricants could lead to suboptimal performance or equipment damage. Fusion will use a contract toll-blender to produce and package lubricants. Initial beta testing with a local aerospace manufacturer and an automobile company has shown substantial improvements in material removal rates, tool life, and part finish quality.

Competitive advantages of Fusion technology are better cooling and higher lubricity than existing metal working fluids; much lower environmental footprint; lower overall system costs than current MWFs; elimination of health risks to workers arising from chemicals and microorganisms; and on-demand coolant and lubricant customization. In recent years, there have been several attempts to improve MWF performance - three most common approaches have been dry machining, cryogenic cooling, and MQL. Dry machining is costly both in terms of tool wear and processing speed. Cryogenic cooling typically uses materials such as liquid nitrogen and is financially and environmentally expensive. Some cryogenic products have had limited success due to its cost and performance limitations - while cooling was exceptional, lubricity was limited.

MQL shows the most promise as a replacement technology for current MWFs and is probably the closest competitor to CHiP Lube. This technology is relatively new and therefore faces technical and economic barriers to implementation. It requires new delivery systems dissimilar to those used for aqueous systems. Pilot-scale implementation has shown that MQL systems are less expensive to operate but require significant upfront investments in both infrastructure and expertise. Sometimes they fail due to poor process understanding. Unless environmental and health costs are considered, many companies believe these costs too high to warrant switching from current systems. On the other hand, a large American automotive company is making significant investments in MQL due to cost savings and higher performance, as well as for environmental and worker health reasons. To date, MQL has only been adopted on a limited basis because it cannot provide the cooling needed in challenging machining operations involving advanced materials.

The successful commercialization of the CHiP Lube system will improve the environment and the health and safety of machinists by removing millions of gallons of hazardous material from the manufacturing workplace. In addition to the substantial environmental benefits for the U.S. and its workers, this system will improve U.S. manufacturing competitiveness. It enables large cost-savings by substantially reducing tool wear, by increasing throughput by allowing higher metal removal rates, and by eliminating the need for management and disposal costs associated with today's legacy MWF products. It will also reduce reliance on foreign oil sources for lubricants due to the use of locally grown soy oil, one of the primary constituents of the proprietary lubricant.

NSF Commercialization Assistance and Impact

The NSF Program Director recommended Fusion Coolant Systems to IA in 2011. Following the introduction, IA worked with the startup to understand the company's short- and long-term goals and needs and found Fusion a good fit with IA's mission and capabilities. Together, IA and Fusion identified the company's need for customer introductions and fundraising. Accordingly, IA worked diligently with the firm to develop a strategy on how best to access customers and partners. It shared advice and feedback obtained from domain experts on the technology and market fit, introduced the CHiP Lube system to potential customers, strategic partners, and acquirers, and facilitated introductions to regional investors. Additionally, IA identified appropriate industry hubs by geography, and advised the company on the right kind of regional investors to approach. IA and Fusion continue to seek and establish strong and highly relevant engagements with individuals and companies in order that Fusion achieves commercial success. The plan is to secure investment capital from the private sector; help the company apply for NSF SBIR supplemental matching funds; increase the company's presence in the market; and identify key human capital.

In 2011, IA featured Fusion as a case study at Carnegie-Mellon University (CMU) during the "IA @ CMU" event to bring attention to the NSF SBIR process and to foster innovation outside of the traditional "innovation hubs" of Boston and Silicon Valley. As a direct result of this event, three CMU-based startups submitted NSF SBIR proposals.

IA activities and outcomes are summarized in the following Table.

IA Activity	Outcomes
Startup Overall Engagement	18-24 months; 20+ hours of interaction; 25+ e-mails sent to and on behalf of Fusion
Introduce Potential Customers/Partners	Introductions to Siemens, Boeing, Behlen MFG, and Distefano MFG; ongoing conversations with these companies helped identify key industries that would most benefit from this company's technology and provide most value.
Feature at Trade Shows and Special Events	Fusion featured as NSF SBIR/IA case study at Carnegie-Mellon University

Where was this startup at the end of federal government funding?

In December 2012, Fusion closed an equity funding round of approximately $0.6M that was enabled by the NSF Phase II grant, and sourced from angel Investors, the Frankel Fund, Michigan Pre-Seed Fund, Enterprise Detroit, and the First Step Fund. Initial aluminum machining tests were completed during the first half of the NSF Phase II project. The results were significant enough to warrant testing for aluminum rolling. These tests were completed in June at the customer manufacturing facility in Lancaster, Pennsylvania. The testing was limited in scope but provided a considerable amount of data, which validated the opportunity and proved feasibility. From this test it was determined that the most appropriate initial application would be in foil rolling. These tests took place in July of 2013. The tests were a success and additional heat transfer tests were completed in November of 2013 to further evaluate the business case for next steps and eventually full- scale implementation.

The relationship with a packaging solutions provider, from both a technical and commercial point of view, is much more advanced than the relationship with the aluminum company. Full-scale, full-speed, continuous runs of cans have been produced using CHiP lube to the point where the system reaches thermal equilibrium. This validates that a continuous, dry can making process can be enabled using supercritical carbon dioxide-based metalworking fluids. Some new technical issues were identified, and a research plan has been put in place with this partner's engineering team input to address these issues. The next scheduled testing date at the customer facility is scheduled for November 2014, at which time a new internally cooled punch/ram assembly will be tested. In 2012, Fusion and this partner entered into a Joint Development Agreement. The agreement does not give this partner any rights to the background IP of Fusion. Management of both companies have begun discussion of a sales/licensing agreement whereby this partner will offer the CHiP Lube dry can forming as a factory option integrated into new machines. Depending on the final technical implementation of the system, there is also the possibility of a cost-effective retrofit solution for existing machines. Testing of a can body-maker with a fully integrated CHiP Lube system has been delayed due to the considerable engineering challenges and manufacturing lead times. The project is still underway with an anticipated first testing cycle in late 2014.

Fusion has the following products available in the market:

1. Fusion Coolant C500 – Cooling only unit.

2. Fusion Coolant L500 – general purpose product providing variable coolant and lubricant control.

3. Fusion Coolant L1500 – higher capacity version of the L500 developed for rolling applications.

4. Fusion Coolant OEM unit – adaptations of the above products to allow their integration into OEM products in kit form.

Total revenues are approximately $0.25M for contract development and product trials. The number of employees at the start of the federal funding was 1.5 and at the end of the funding was 2.5. The technology developed through the NSF funded R&D has broad applicability to solving customer pain points across a range of applications. The company is currently in the process of raising additional outside funding to continue operations and growth. Products are customer ready and have been reliably demonstrated in long-term operation at customer premises. Fusion is engaged in several ongoing product trials with customers in both the aerospace and automotive industries.

The need for two levels of product has been demonstrated, one with a low-volume lubricant delivery capability and the other with a high-volume lubricant delivery capability. This two-tiered approach allows Fusion to attack a broader range of applications in the forming/rolling area. Updated hardware was designed, fabricated, and delivered from their vendor during the project period. During the final period of the NSF grant some testing of alternative lubricant products was completed but little progress was made on in-house development of custom lubricants. During the reporting period, Fusion management has developed a strategic relationship with a local specialty chemicals company that operates as a toll manufacturer and has considerable expertise in the development of custom lubricants for metalworking applications. In the future, it is likely that Fusion will continue its custom lubricant development efforts through collaboration with this strategic partner.

Where's the startup now?

Today's metalworking fluids compromise cooling for lubrication. Add oil and you reduce cooling. Add water and you reduce lubricity. Oil-in-air minimum quantity lubrication can lubricate but does not cool well. Liquid nitrogen can cool but does not lubricate well. Only Fusion's supercritical carbon dioxide gives maximum cooling and lubrication potential at the same time, increasing productivity and reducing system-level costs. Fusion's supercritical solution flows to the point of machining as a single-phase system and is released from high pressure, producing a strong cooling effect, and delivering dry, or enhanced lubrication.

Fusion's critical patents secure the process of using supercritical carbon dioxide as a machining coolant and vehicle to deliver dry and enhanced lubrication to a cutting zone for applications demanding high lubrication and high cooling. Cooling and lubrication are achieved via single channel delivery because supercritical carbon dioxide completely dissolves lubricants. This allows for unprecedented heat removal potential and lubricity. Compelling results have shown that Fusion's inventive process offers a multitude of cost saving benefits for less operating capital than conventional cutting fluids. Fusion's supercritical carbon dioxide technology promotes a healthy work environment because it is bacteria free, often eliminates post-machine cleaning steps and has minimal environmental impact, as shown by published research. This disruptive technology has been proven in aerospace, bio-medical, automotive, and heavy equipment manufacturing applications that were previously considered out of reach for MQL technology.

Why should a customer use this technology? Supercritical carbon dioxide flows to the point of machining as a single-phase system and is released at high pressure. The high pressure produces a strong cooling effect and delivers dry or enhanced liquid lubrication. In either lubrication mode, dry or enhanced liquid, clean and dry chips are produced. Fusion's process does not require re-circulation or disposal of the metal working fluid. Supercritical carbon dioxide is an excellent coolant when released from high pressure, a lubricant which on its own is as effective as a semi-synthetic metal working fluid. It is also an excellent solvent which can, in enhanced mode, provide straight-oil levels of lubrication. It also allows higher cooling and higher lubricity with higher pressure, leading to higher productivity. It does not clog and lowers operational costs.

Case Study Questions

1. Discuss the role of technical founder(s) in a university spin-off company: The art of control
2. Team building – how difficult is it in the manufacturing space? Does startup location matter?
3. Fundraising challenges when developing technology meant for traditional industries such as manufacturing
4. Address the question of long sales cycles
5. Right business model?
6. Identify areas where startup needs help
7. Growth and exit strategy
8. Clarity of communications, to align expectations with strategic partners
9. Take-away(s)

7 A Sense of Touch

Technology Sector: Displays
Startup: NextInput, Inc.
Website: www.nextinput.com
Location: Atlanta, Georgia

Federal Funding Timeframe: January 2013 – March 2017
Funding Amount: $0.77M

This narrative is a story of: Need for private equity funds beyond federal funding; scale-up/production challenges; founders/team issues – bringing in outside CEO; original location/ecosystem not too conducive

Startup Formation

NextInput, Inc. ("NextInput") was incorporated in 2012 by Ian Campbell and Ryan Diestelhorst, the two founders of the company who met at the Georgia Institute of Technology ("Georgia Tech") while Ian was pursuing a dual master's in aerospace engineering and an MBA and while Ryan was working to obtain a doctorate in electrical engineering. In 2008, Ian completed his masters' degrees and began work as a management consultant, while Ryan continued to pursue his Ph.D. in analog integrated circuit design. The friends stayed in touch, however, and often brainstormed around new technology innovations and entrepreneurial opportunities related to electronics, sensors, and touch interfaces.

The two founders soon realized that highly sensitive micro-electro-mechanical systems (MEMS) sensors could be applied to user interfaces, specifically in consumer applications involving touch. The pair tinkered with pressure- and force-sensing touch systems until 2011 when they reached their first critical milestone – the development of a touch interface prototype with the ability to recognize high-resolution, high-accuracy force thus enabling a third dimension of touch to complement standard two-dimensional touch interfaces. They demonstrated an early three-dimensional (3D) ForceTouch prototype to Atlanta-based investors who were intrigued and provided valuable feedback. At the time, touch technology markets were exploding, expanding beyond the massive mobile application markets to the automotive, home automation, wearables, and the internet of things (IoT) sectors.

Ian convinced his wife that it was the right decision to quit a well-paying management consulting job and focus his efforts on building a company based around 3D touch innovation, adding that he would have to forego a regular paycheck for six months. It turned out to be eighteen months before he could pay himself a regular salary. Fortunately for Ian, his patient wife continued her professional work during this period bringing home a steady paycheck. Using personal resources and family members, Ian and Ryan obtained investments worth $250K. Ryan completed his Ph.D. in December 2012 after a few months of working at NextInput during the day and writing his dissertation at night. For him, it was a relatively easy transition from graduate school to the startup lifestyle since he was already used to living on a graduate student's salary.

Technology

In 2012, force sensors already existed in the marketplace but were for the most part, a solution looking for a problem. These types of sensors were then not specifically tailored to the touch-interface market, although many consumer electronics firms, such as smartphone original equipment manufacturers (OEMs), were considering adding force sensors to their products. MEMS sensor technologies typically address the high-volume, low-cost, consumer market where it is necessary to go-to-market with the right functional specifications. Feedback provided by advisors, investors, and domain experts convinced the founders that their technology had great commercial potential[13].

Neither had MEMS design experience, so they taught themselves how to design these types of sensors by connecting to experts like Professor Oliver Brand at Georgia Tech and Steve Nasiri, founder of InvenSense, the largest MEMS startup company to go public. The team spent a year evaluating existing sensors in the market, building, and improving ForceTouch prototypes, and meeting with investors to raise additional capital. The growing team had many debates on the technology development path to pursue, and received well-meaning advice from technologists and business strategists from the Advanced Technology Development Center (ATDC) at Georgia Tech.

University researchers such as Professor Brand, Interim Director of the Institute for Electronics and Nanotechnology (IEN) at Georgia Tech, helped to conceptualize early force sensor designs and provided access to cleanroom facilities at IEN that NextInput used to prototype two generations of its force sensors. MEMS industry pioneer and technology visionary Dr. Kurt Petersen, the driving force behind six MEMS startups, provided valuable advice early in the company's history. The founders wrote their own patents for sensor design and acquired others with similar designs from independent inventors to hedge against competition. The company has six key pending patents in this space for mechanical design, firmware, and system solutions. An intellectual property (IP) law firm helped with their IP strategy, and they obtained critical advice from entrepreneurs like Steve saying – "Only file what is applicable and readily relevant to your technology".

The two founders greatly enjoy living in Atlanta and owe a great deal to Georgia Tech, the IEN, ATDC, and the local entrepreneurial community for early support; however, they felt the strong pull of Silicon Valley. While Georgia Tech is a great research hub for sensor science and for early technology development, comparable to MIT and Stanford, the entrepreneurial ecosystem in Atlanta lacks sufficient market knowledge and insights around hardware technologies and hardware startups like NextInput. For later stages of technology development and for company growth, Silicon Valley has the expertise and the required supply-chain infrastructure to support a growing company.

In early 2014, NextInput formed a strategic partnership with market-leading contract manufacturer Flextronics' Lab IX, a hardware accelerator in Silicon Valley. The company began working from this facility on sensor qualification and ForceTouch solution commercialization in mid-2014. Now it is building a base of operations with several employees in San Jose. The founders made a key decision to manufacture ForceTouch sensors in Taiwan after considering the U.S., Germany, Canada, and Sweden. Germany has a deep legacy in MEMS-related manufacturing, but the cost structure was not suitable for consumer applications. The team found similar conditions in the U.S., Sweden, and Canada. Taiwan was attractive due to its decades-long experience in the MEMS industry, modernized production facilities, strategic location near key manufacturing centers, supply-chain partners and OEMs in Asia, and a reasonable cost structure.

The founders engaged the same foundry used by InvenSense - Steve leveraged his original relationships to ease transition of the startup's sensor commercialization from a university lab to a contract MEMS foundry. The risk of improper technology diffusion was low – in fact, protection of IP is a key element of the fabless semiconductor model, and this foundry prides itself on allowing no exceptions. This facility is a smaller pure-play MEMS contract foundry with revenues of approximately $30M; however, with small size comes agility. The foundry can perform minor process development to help NextInput enter smaller markets, assist with early design and process validation for low- to medium-volume production, and improve the startup's bargaining position when the company seeks additional equity financing.

The smartphone market, NextInput's goal, will require much larger foundries capable of billions of units per year in output. The latest ForceTouch sensor design and process have been finalized and mass production for early market entry customers will begin in 2015. The firm plans to refine its sensor processing and optimize the sensor supply-chain in 2015, and in early 2016 scale up to large-volume sensor and solution manufacturing. The full solution – sensor, mechanical system, software, and firmware – will be available in 2015. Currently, Ryan and Ian spend considerable time in Taiwan, deeply involved in process design, rapid iterations of manufacturing processes, and building up their supply-chain.

Customer Discovery and Business Model

The best kinds of customers for a hardware startup are early adopters willing to fund technology development. A few of these were serendipitously obtained. In mid-2012, the founders were attending a tech conference in Atlanta, when a local IP attorney walked by their booth, looked at their early prototype touch system, made a few connections in his mind, and later made a very important introduction to his client, a Tier-1 automotive supplier. In the automotive industry, serious pain-points exist for touch systems – for example, the need to sense touch forces correctly and accurately without generating false positives that could result in accidental activation of critical subsystems such as cruise control. The team met with this supplier, discussed the pain-points for automotive touch systems, and mapped out a plan to solve them using ForceTouch technology. This relationship proved critical on multiple fronts – it led to more non-recurring engineering (NRE) projects which helped reduce the need for too much initial capital, it allowed the company to refine its technology to satisfy specific customer requirements, and later would lead to a key investment.

The progress and development success with the automotive partner also led to other NRE-type opportunities with joint-development partners in multiple industries. However, the company soon realized that such NRE work was consuming too much of the company's resources. It is therefore slowly winding down this type of work to just a few strategic partners and focusing on core ForceTouch technology and product development. The company has also narrowed its market entry strategy. Although NRE contracts helped to successfully develop complex solutions for automotive touch zones, these manufacturers expect critical components like force sensitive touch zones to satisfy often difficult and rigorous requirements.

With sensor technology development nearly complete, the company's focus has now shifted from specific customers to entire markets - by developing overall market solutions for many different applications such as active styluses, personal navigation devices, gaming, IoT, and wearables – by which the company can generate the highest return for its development dollars. In early 2015, the company plans to release a set of development kits and application notes for these early market entry points. As a hardware startup in the sensor space, the smartphone market is the holy grail, but it would be a fallacy to focus exclusively on this market at this early stage. Instead, NextInput's strategy for the consumer market has been to enter segments with simple force-sensing solutions requiring lower volumes but higher Average Sales Price (ASP), begin to replace the $4B medium-volume resistive touch-screen market, used in applications like point-of-sale systems, GPS devices, and automobiles, and eventually augment capacitive touch in high-volume smartphones by adding in force, the third dimension of touch.

This strategy will establish and refine early production runs for applications that require low volumes, can support a larger ASP and allow the company to earn healthy margins along its scaling path as volume increases and ASP decreases. The company will then scale through tier-1 component suppliers that manufacture display modules for smartphones, and ultimately OEMs themselves. To properly determine the point in the supply-chain where the firm needs to operate is why it needs to be headquartered in Silicon Valley.

The business model has also evolved, as it does for most successful technology startups. Significant assumptions made initially turned out to be incorrect or different – for example, the idea that both sensors and software can be sold proved to be challenging. Veterans with domain expertise know better: a company cannot charge customers for software or firmware in the high-volume consumer market, but perhaps in other markets like the automotive industry. The business model now simply calls for sensor sales - not directly to an OEM like Samsung or even a contract manufacturer like Foxconn - but deeper into the supply-chain, perhaps to a Tier-1 display vendor.

NextInput made many early mistakes pursuing large customers in the smartphone space without a fully baked product and becoming distracted from core product development. Sub-optimal solutions disappointed customers so the focus became one of technology development. In 2013, the company realized that if they continued "chasing dragons" in the smartphone world, they would increase their risk of going out of business without an excellent product, and simultaneously have unhappy NRE customers. Funding from the National Science Foundation (NSF) Small Business Innovation Research (SBIR) program proved critical. It helped the company develop its core product and allowed the team to focus on what really mattered – bringing the technology to market – rather than chasing NRE contracts.

Business models for startups in the user interface (UI) and touch technology space can be generally segmented into three categories – improve existing resistive and capacitive touch technology; improve software for touch technology; and introduce new UI features and functionality. Many startups in this space will be successful because the market is so large. Still, the better segments for startups in this market are the latter two. In the case of software, it would necessarily be licensing plays – where the risks are low, and the startup can be profitable, but revenue and profit do not scale well. A hardware sensor company, on the other hand, can go public. NextInput's business model is more related to the latter two categories – improving existing touch applications and adding new 3D touch functionality to enable a richer UI experience.

Financing

The startup's initial funds began with Ian's early investments and later an additional $250K from friends and family members. Raising money from close friends and family members can be a weighty decision – the founders would be despondent if any of its investors lost money but would be especially saddened if these were family and friends. Ian's personal and professional track-record and that of his partner, Ryan, convinced early investors. Making the company successful for early investors is a top priority.

Ian and Ryan were the first true believers. Now, having seen great progress, initial investors are believers as well. In fact, some of them participated in the second seed round in 2012 for additional funding of $925K from family members, and angels in Atlanta, the Midwest, and the Bay Area. In 2013, Steve Nasiri officially became an advisor to the company, and later led a new financing round in mid-2014. As a lead angel, Steve helped pitch the company to Silicon Valley's Band of Angels, a well-recognized and well-organized angel group. There was little need to educate this group on the potential of a MEMS opportunity – many in the group had become successful in the MEMS space both as entrepreneurs and investors.

The two founders have an equal equity stake in NextInput. The first six employees of the company have stock options and some restricted stocks have also been offered to Board members and the five advisors. Ryan and Ian together still hold a majority stake and hope to maintain it through the Series A round of about $5M slated to close in mid-2015. This would provide adequate runway to profitability in 2016. By then, it is expected that the company will have completed its relocation to Silicon Valley. The two founders were introduced to the NSF SBIR program by Georgia Tech's ATDC – this center provided help in crafting the first grant proposals. So far, the company has been awarded two NSF SBIR Phase I grants for $150K each and one Phase II grant for $500K. The two founders believe that the value of this non-dilutive capital was well worth the effort of writing the grant proposals, although they recommend more transparency in communicating where the proposal is in the grant process - hiring decisions and capital equipment purchases were all dependent on knowing when funds would be disbursed.

Ryan maintains valuation is always a tricky subject with investors, but if the company is achieving its milestones and has several financing possibilities, a mutually agreeable valuation can always be found

"It's a kind of dance. There is no correct valuation formula, but we believe it should always be milestone-based. For example, achieving a qualified sensor and manufacturing process, achieving production quality solutions, or obtaining the first purchase order from a customer– as we accomplish these milestones, we can command a higher valuation."

"We really never stop fundraising," added Ian. "We keep investors updated with one or two milestone events every quarter. There are also cash-flow considerations. It's good to have the maximum possible amount of cash in the bank, accounts receivable, and NRE backlog as a cushion. That helps with investor confidence. They see that the company is adequately capitalized and is earning sufficient NRE or grant revenue to lengthen the runway. NSF funding is also critical. The funding we've received from NSF to-date has played a huge part in minimizing our financing hurdle. That money was directly leveraged to hire engineers, purchase capital equipment, and fund sensor commercialization efforts. NSF funding will also be a large part of our 2015 financing plans."

"NSF has helped our company in other ways as well," adds Ryan. "For several years, the NSF has funded the National Nanotechnology Infrastructure Network (NNIN), of which the Georgia Tech IEN is a member. Most of the equipment we've used in the Georgia Tech cleanrooms was either funded directly or subsidized by the NSF NNIN, and that's really been an enabling resource for us."

Company Growth

Currently the company has six full-time employees in Atlanta and San Jose. There are five board members – Ian, Ryan, one board member from the first seed-round, one from the 2012 angel round, and Steve Nasiri, representing the Band of Angels in the third round. Ian relocated to Silicon Valley in mid-2014 and aims to fill out the operations, finance, marketing, and sales team with Bay Area hires and some from the executive ranks at InvenSense, including a VP of Operations who helped build that company's later-stage supply-chain. Ryan and key technical employees will remain in Atlanta until mid-2015 where R&D work will continue to further reduce technology risk. The team continues to use the Georgia Tech IEN for metrology and the development of quality assurance equipment such as wafer-level probe stations. Currently, Ian leads sales and business development efforts in the U.S., Japan, Korea, Taiwan, and China. However, the company plans to hire a Director of Sales in early 2015 to set up global sales teams.

Ian was asked about his evolving role as the CEO of the startup, first as a technology developer, then as primary fundraiser, and currently as head of sales.

"The role of a CEO in a startup company changes as the company grows. If you look at most MEMS or semiconductor startups, you will find that the founder-CEO usually gets the company through initial R&D to a scaling point. When the company transitions to an operational, supply-chain-oriented, and sales-heavy organization, it would require a new CEO. Then again, the CEO who scales the company often doesn't lead the company through an exit, or stick around much after the exit. My role now is to build the best possible foundation for NextInput. We're focused on the most capital efficient means to get to our next immediate milestones to maximize our valuation, which in turn maximizes the unrealized return-on-investment for our current investors. Every decision we make today impacts our future valuation, and that includes who is in the driver's seat. If it's the best thing for the company to bring in an outside CEO at any point, we'll do it."

According to the founders, mistakes made hiring people led to the most painful lessons.

"The first "hire" was the founder partner. It is critical to have an excellent co-founder that is equally committed, equally skilled, and whose skills complement your own. Ryan and I make a good founding team. Ryan brought strong technical expertise and realism to complement my business strategy skills and optimism."

"Other early hires are just as important. We made some mistakes hiring early on. As relatively novice business owners, we were naturally very easy to impress. We brought on people who had great resumes but were missing the intangibles – the strong desire to be a part of a team doing something new, and to put in the long hours and suffer the slings and arrows of bringing that new thing to the market."

"Now the company focuses on hires that fit within the company's entrepreneurial culture. Startup culture is always a bit ad hoc, and usually reflects the personalities of the founders. We don't want to force a 'NextInput Culture' on our team members - we want to avoid making rules at this stage. Our emphasis is more on building camaraderie and a strong work ethic – a team in the trenches working together, getting over hurdles, achieving milestones, and building ownership in the company. I like to think that Ryan and I lead by example – what we do and how we do it are examples to the rest of our team."

"Accomplishing something substantial is more important to us than money ever will be. It's a legacy thing. Ian and I want to be masters of our own destiny, which is something very few people fully achieve in life. Right now, we are beholden to investors, as we should be. But we strive to be independent like many of the successful entrepreneurs that we model our careers after – guys like Steve Nasiri, who don't necessarily have to work another day in their lives, but still pursue opportunities for personal fulfillment. We wouldn't get too far without those types of thought leaders to lean on and to learn from."

When asked what lessons have been learned from Steve, Ian answers

"Steve is a classic hardware entrepreneur: detail-oriented, an excellent engineer, and an excellent strategist. He can go 'up and down the ladder': able to tackle the highest corporate strategy problems at the top, but also able to solve nuts-and-bolts engineering problems at the bottom. Ultimately, it was his ability to recruit and motivate an excellent team at InvenSense that allowed him to execute his vision. We try to emulate his skills and approach – technology education is continuous, there's always a solution, and if you're smart enough and work hard enough, you'll reach the end goal."

"Like any hardware startup company, we're planning on an exit in three to five years, no matter what year it is," jokes Ian, before striking a more serious tone.

"Our most likely exit path is through an acquisition by a key partner, customer, or competitor. Sometimes we dream about an IPO, and if we stay focused on achieving our milestones, perhaps we can get there. But we will never let our ambitions run away with us to the point that we lose focus on generating the best exit for our shareholders."

Key Relationships

For market insights, the founders rely on their own research and the wealth of knowledge they glean from their Board of Directors and Board of Advisors. Bringing a new hardware technology to market requires three things – a startup with a new technology, a customer that wants the technology, and ecosystem partners to help solve manufacturability and commercialization challenges, like contract foundries, contract manufacturers, component vendors, and other technology enablers.

"We need to maintain and grow all the business relationships we have now, and cultivate new partnerships to enter key markets, like the mobile or the automotive touch system markets. Our initial focus was on selecting a good foundry partner to produce our ForceTouch sensor. Steve helped us form a relationship with our Taiwanese foundry partner. The second set of partnerships that will be important for us are solution partners: tier-1 vendors or contract manufacturers that can help us take our sensor, package it into a sub-assembly solution, and sell the solution to OEMs, our end customer that will productize ForceTouch solutions."

The founders have recruited key people to join their Board – the directors and advisors have opened numerous doors with key introductions to their vast networks. The company has also done plenty of their own legwork to build pivotal business relationships.

"We're working with market leaders in the mobile, laptop, automotive, and general consumer electronics sectors on ForceTouch applications. Our goal is to have numerous pathways to product revenue in 2015."

When asked if trade groups and trade shows were important for hardware startups –

"I believe they are, but it completely depends on the stage of the company. We have attended and spoken at many trade shows and technology award events and have brought home trophies. That's great validation of what we're accomplishing, but ultimately, we're about building a real business. For some tradeshows like CES, it's easy to just be 'background noise'. We prefer smaller events like the NSF SBIR Phase II event in Baltimore this year, or the recent MEMS Industry Group conference on Wearables Sensor Technology in Santa Clara. Those venues give us a chance to engage directly with engineers and product developers seeking new touch solutions, understand their issues, and help sell the value of ForceTouch."

Startup Experience Summary

Ian and Ryan have experienced much of the ups-and-downs of the startup world and are hoping to soon experience the reward – a profitable exit. Through it all, they've learned a great deal about what it means to become an entrepreneur, a one-way street from which they might not return.

"I can't imagine going back to the corporate world," says Ryan. "I was a student intern at one of the big semiconductors manufacturers and got a brief glimpse of the life of an engineer solving small but important pieces of a much larger industry shift to 45 nanometer semiconductor manufacturing. I was struck by how narrow a picture each engineer had about their contribution – many had no idea why they did what they did, although they didn't seem to mind. I also interned at the Jet Propulsion Lab (JPL) after receiving my Ph.D. and had a splendid job waiting for me there at a world-recognized national lab – a Ph.D. student's dream – but I believe I made the right decision. The startup experience has taught me so much in such a short amount of time; I have no regrets."

Ryan had trouble explaining what it was exactly that made him want to be an entrepreneur.

"I don't know exactly. Most of my traits are not consistent with the typical entrepreneur – I can be a bit pessimistic and overly cautious. I don't enjoy the startup social culture or business development or sales all that much. That's why I need a guy like Ian! I'm just an engineer that prefers to work on problems that have a clear and immediate impact on the direction of the organization. It's much more exciting for me that way."

He had less trouble explaining the significant progress NextInput has made in his characteristic nonchalant tone.

"I think we've made pretty good progress, but I am always tense, looking at the path forward, worried about money, people, technology development, competition. I think we both misjudged just how difficult a hardware startup would be. We didn't have a complete picture – and few first-time entrepreneurs do. At this point, we thought we'd be further along, but mistakes we made cost time and money. Entrepreneurs make many wrong decisions, but you hope the right ones outweigh the wrong ones."

Ryan added some advice for a young entrepreneur considering a startup in the MEMS or semiconductor space.

"First, try to know what you know and what you don't know – seek out good advisors who do know the space and absorb information from them. Don't worry too much about very long-term goals, focus on the next milestone and execute. Don't spend too much time and money on IP protection. Build and demonstrate the first prototype and applicable IP will naturally result. Fill gaps in your team with like-minded, entrepreneurial people, but with complementary skill sets."

Ian mostly agrees with Ryan.

"Assume you don't know much about what you are trying to do. Stagger your meetings with advisors, investors, and partners – go after the most important ones last. You don't want to come across as clueless in your first meeting, you need to work your way up to the Nasiri-types. If we had talked to Steve too early, he would have laughed us off. Keep short-term focus on the next proof-points to achieve, prove the idea to a solid cofounder first, then prove some early concept to an angel investor, then prove a technical solution to a customer, then prove manufacturability and a commercialization path to a later-stage venture capital firm. We are still not done with our proof-points, but that's how we track our progress and what we need to do next."

When Ian was asked about his strengths and weaknesses, he had this to say –

"I think I'm good at putting the right people together – finding what the team and company is missing and filling those gaps. My primary weakness has always been that I'm a bit too optimistic. That's why I think Ryan is a great complement – he's a very practical, realistic guy that helps keep me grounded. I completely trust Ryan and we work well together – we were friends for years before we started NextInput, and we'll be friends long after the company exits."

Where was this startup at the end of federal government funding?

Equity investment rounds of $8M in December 2015 from Sierra Ventures, GoerTek Group, and Intel Capital; and an additional $5M in June 2016 from Danhua Capital, Cota Capital, and UMC Ventures were successfully closed. NextInput has the following products in production for various customers and markets:

- FT-4100 - Consumer qualified flat sensor.

- FT-4500 - Automotive qualified sensor based on the FT-4100.

- FT-5100 - Sensor with a mesa for applications that require large mechanical tolerances in the assembly.

The company has the following issued patents:

1. Microelectromechanical Load Sensor and Methods of Manufacturing the Same Patent No: 9,032,818, UNITED STATES Application Date: 07/03/2013 Date Issued: 05/19/2015 Status: Granted

2. Miniaturized and Ruggedized Wafer Level MEMS Force Sensors Patent No: 10296-028US1, UNITED STATES Application Date: 07/12/2016 Status: Pending

3. Ruggedized MEMS Force Sensor Having a Tolerance Trench Patent No: 10296-034US1, UNITED STATES Application Date: 06/10/2016 Status: Pending

Total revenues as of March 2017 was approximately $3M. The number of employees at the start of federal funding was 6 and at the end of the funding period was 28. In 2015, Apple created the ForceTouch market with the release of a full line of force-enabled products, including the Apple Watch and Macbook Pro. Most notable of Apple's offerings was the iPhone 6s, released in September 2015, which integrated 3D touch into the iOS user interface. Since then, Apple has continued to double down on force-enabled products and more deeply embed force into every level of user interaction in the iPhone. While this development created plenty of excitement in the market and allowed NextInput to raise significant capital, it was not immediately followed by a response from Google. To date, Google has yet to integrate force into the Android user interface, which is required to open a market for NextInput in mobile screens. In response to this development, NextInput has pivoted to button replacement applications to reduce time to revenue while Google determines when and how it will integrate force into Android. This pivot still includes mobile, and thus has the potential for significant volumes. In fact, NextInput's first major design win was with a large mobile device manufacturer in 2017. The mobile market is wide open for the replacement of dome switches under both home and side buttons to allow waterproofing and edge-to-edge displays. NextInput's force sensor is an ideal technology for this application and has allowed it to gain significant traction in 2017.

Where's the startup now?

NextInput, Inc., headquartered in Silicon Valley, provides MEMS-based force-sensing solutions for touch-enabled devices in consumer, wearable, IoT, automotive, industrial, medical and other applications. NextInput is pursuing two additional markets to complement the fast, high-risk revenue environment that mobile presents:

Automotive: The world's largest manufacturer of automotive steering wheels (over 30% market share) is engaged with NextInput to develop touch interfaces for steering wheels with over $1M in executed R&D contracts. Work has advanced to customer projects with major automotive OEMs in Europe and the U.S., and AEC-Q100 qualification has been completed to deliver product to these customers.

Industrial: Multiple major European brands have engaged with NextInput to implement force sensitive buttons and x-y screens for industrial applications, where capacitive and resistive are ripe for displacement due to limited reliability and functionality.

NextInput currently has design wins with one mobile handset manufacturer, which it will begin fulfilling orders for in Q3 2017, and a Tier-1 automotive supplier, who will begin ramping volume production in 2018 with a limited subset of high-end OEMs. It has a few smaller design wins in consumer spaces such as stylus and toys, which will also begin shipping in late-2017. The end of 2015 brought new investment to NextInput in the form of a Series A round of $8M. Series A1 quickly followed in 2016 to accommodate investors that were unable to participate in Series A due to a cap on the round size dictated by NextInput. Total institutional investment now totals $13M. NextInput will continue to raise capital through 2018 to fuel a ramp-up of customer/market coverage and the underlying production capacity necessary to deliver high-quality products with rapid lead times. Over the next several months, the company will expand its world-class team and increase production capacity in anticipation of customer production ramps. NextInput will also continue to innovate by developing the next generation of MEMS-based ForceTouch® solutions.

"Successful completion of Series A is a major milestone, and I am pleased to have high caliber investors such as Sierra Ventures, Intel Capital and GoerTek Group Co., Ltd. backing the company," says Ali Foughi, NextInput CEO and Founder. "The market is ripe for the adoption of force-based 3D touch, for which NextInput has the best solution. The funding will allow us to scale the company and build significant value for our customers, employees, and shareholders." "We have confidence in NextInput's team and are thrilled to lead their Series A round," says Ben Yu of Sierra Ventures. "We believe NextInput's innovative solutions will disrupt the multi-billion-unit touch market."

Case Study Questions

1. Comment on building a hardware company
2. Assess commercial potential and value proposition
3. Develop a business strategy/business model for the company
4. Lessons learned about building the right team
5. The importance of geographic location
6. Comment on hollowed-out U.S. supply-chains for certain technologies
7. Identify risk elements (technical, team, market, finance) and how to manage/mitigate them
8. Additional funding strategy
9. Barriers to entry
10. Growth and exit strategy
11. Start-up valuation exercise
12. Take-away(s)

8   What I Wear Harvests Energy

Technology Sector: Wearables
Startup: Perpetua Power Source, Inc.
Website: www.perpetuapower.com
Location: Corvallis, Oregon

Federal Funding Timeframe: January 2010 – March 2015
Funding Amount: $1.11M

This narrative is a story of: The need for private equity funds beyond federal funding; sales/ bootstrap growth; IP play (over thirty patents); industry partners; scale-up/production challenges; contracts and joint development; team consists of seasoned veterans from corporate America; spin-offs and acquisition of small companies for synergy/ growth

Startup Formation

Jon Hofmeister, President of Perpetua Power Source, Inc. ("Perpetua"), founded the company in 2005. Nick Fowler, the Executive Chairman, joined the company in 2008. Jon's background consists of both technical and business experience. He grew up in Oregon and obtained dual degrees in civil engineering and engineering geology from Stanford University. One of his civil engineering professors, an expert in risk analysis, started a company and asked Jon to join her. He agreed, for it was a nice fit with his skill set, and exciting to be given an opportunity to make a significant impact. In less than eighteen months, the company grew to over fifty employees and was acquired. After the acquisition, Jon continued as a consultant to the acquiring company for five years. During this period, he completed a master's degree in civil engineering from the University of Washington. Subsequently, Jon worked as an engineer for URS Corporation, and then was employed by the State of Oregon. He also obtained a professional engineering license.

Looking to broaden his skills in business, Jon enrolled in the Master of Business Administration (MBA) program at the University of Oregon while working as an engineer. As part of the program, he researched possible business opportunities, including those involving thermoelectric technologies. Jon graduated early with an MBA and took a leave of absence from his engineering work to travel to the London School of Business and Finance to further incubate his ideas and to eventually decide whether to pursue thermoelectric technology. When Jon returned to start Perpetua in late 2005, two partners, Paul McClelland, and Marshall Field, from Hewlett-Packard's Corvallis-based inkjet printing group joined him. The company obtained its first financing in December 2006.

Nick obtained degrees in economics, industrial engineering, and engineering management from Stanford University. He was invited to join HP straight out of graduate school by Dave Packard himself. HP was large, with many near-independent business units where autonomy, creativity, and an entrepreneurial culture of innovation were cherished. Nick, as a post-graduate Fellow, worked in the CEO's Office of Corporate Strategy that, among other strategic activities, studied opportunities in HP's technology portfolio. Over the years, Nick was exposed to various facets of the company – operations, technology acquisition, technology development, manufacturing, and company acquisition. After twenty-five years, Nick left to try to replicate what he did at HP outside of HP, by finding startups to invest in. Nick insists he is not an idea man, but is good at teasing out early-stage ideas to turn into viable businesses. He felt that Perpetua had a compelling business proposition, and that Jon had assembled a cohesive team of technology veterans and young engineers with a powerful blend of knowledge, energy, and experience. To him, it looked different from other startups. Here was a team and leadership committed to success.

"I offered some experience taking companies to market, and experience in investment, financial and business strategies. I liked the culture of teamwork and frugality, the latter a quality that you seldom see in startups funded by venture capital (VC). I remember when I first visited the company – a warehouse location, no designer furniture; Jon even remarked that all the furniture was from a surplus sale at the local university, and that no desk cost more than a few dollars. The personal chemistry with Jon was apparent, right from the very beginning."

Technology

Jon found thermoelectric technology intriguing. The National Aeronautics and Space Administration (NASA) had deployed it over many decades, finding it extremely reliable over the long-term, in some cases over a period of fifty years. It also has never failed when used as a satellite power source. The technology is robust and not chemically reactive, so components not being subjected to wear and tear last for decades. It is also efficient at high temperatures. One drawback is its high cost. Jon researched work being performed at a Department of Energy (DOE) National Lab for usable power from lower temperature differences along with lower-cost manufacturing approaches[14]. The DOE work looked promising in the lab, but he needed a manufacturing-savvy partner and a cutting-edge approach, like those used in inkjet technology, for the product he was contemplating. He found such expertise in Paul and Marshall. Most of the initial thermoelectric technology was licensed from the DOE and the University of Oregon. A merger with ThermoLife, a potential competitor, brought in additional intellectual property (IP) developed by that company. Its chief scientist, Ingo Stark, now holds the same title at Perpetua.

How do you decide what IP you need and how do you acquire it?

Jon - "One criterion we use is its viability of being a 'platform' technology with opportunities for multiple products and applications. It's harder to attract investors and to recruit talent if the IP is an incremental development of current technology. In looking for an IP set, there must also be an element of market pull."

Nick – "The National Lab technology was platform technology – broad background IP that gave our company its first ability to access product development with which we carved out core geometries. Building on top of that, we used ThermoLife technology to develop adjacent geometries. Technology licensed from HP added a specific instantiation of a product. The team then filled out the framework and put flesh to the bone. Perpetua currently owns roughly thirty patents and pending patents."

Describe the primary applications of this technology.

"The low temperature-difference area of thermoelectric technology, meaning producing useful power at room temperature or other commonly occurring temperatures as opposed to very high temperature applications as in boilers, is very broad. Essentially, wherever there is heat coming off equipment such as motors, pumps, pipes, water lines, etc., coupled with a need for longer-life batteries or renewable power, there is a potential that thermoelectric technology will be useful. The diverse set of applications includes powering security sensors, commercial energy management, automated processes, and infrared sensors. We have, however, chosen to initially focus on two prime areas - powering wireless sensors for industrial process and equipment monitoring; and powering wearable devices."

Jon, would you then say that technology development is done?

"On the industrial products side, the Power Puck™ is done. It has already passed a high bar – it is Intrinsically Safe (IS) certified and connects directly with sensors manufactured by GE, Emerson, Honeywell, and others, along with attachments, accessories, and customer support. On the wearables side, we are in the process of integrating our specialized thermoelectric material into flexible form factors with commercial partners. That work involves adjustments rather than fundamental research – refining the original work for better performance, packaging, and related large-volume manufacturing capability."

Customer Discovery and Business Model

To understand the market and find potential customers involved upfront work using a set of filters - capabilities overlap, market needs, price, manufacturing volume tradeoffs with price point, and should Perpetua take the complete product to market by itself or would it be better to partner with more established players? Early on, the R&D 100 award gave the company immediate visibility and facilitated a helpful customer discovery period. There were also many incoming inquiries after the MIT Technology Review article on Perpetua. Building on the learning from these market interactions, the company consciously architected a sales strategy, identifying the most attractive and most influential companies in the targeted market segments. The company then executed on that strategy through product conversations and disciplined execution from proof-of-concept through market roll-out. On the industrial applications side for wireless sensor networks, Perpetua is undergoing rapid growth. The company continues to hire talent to transition from a technology-development to a market-development company.

"Generally, you want to under-promise and over-deliver. The whole energy harvesting opportunity has been much hyped and we deliberately try to avoid over-exaggerating the possibilities. Our strategy is to demonstrate in controlled environments, then deliver in real-world situations. We did extensive work with the largest sensor manufacturers who are helping sell our Power Pucks. They integrate our product with their sensors, then sell to their customers such as Chevron, BP, Exxon, and Shell. The industrial market is moving forward nicely."

The value proposition for the industrial market is clear: for the cost of a single battery replacement, the Power Puck offers a virtually unlimited power source. If batteries are used, depending on required data rates, lifetimes tend to be just a few months to a year – with Perpetua Pucks lifetimes can be twenty or more years! Typically, frequency of data collection and data rates drive battery life. For example, oil companies such as Chevron are turning to wireless sensors to log parameters such as pressure, temperature, and flow rates, at various sampling intervals. Wireless sensor networks are lower cost, more flexible, and scale much more easily than wired networks once the network is deployed and in place. When higher data collection rates are desired, battery replacement costs to support the higher rates can be prohibitive. Batteries have always been the Achilles heel of wireless sensor networks. Perpetua directly addresses this pain-point and enables the full benefits of wireless technology.

On the industrial side, what are the business model risks?

"In industrial applications, currently over ninety percent of wireless sensor networks use batteries. The rest use solar panels. In heavy industrial applications like oil and gas and power plants, with dusty and dirty environments, maintenance costs are high for solar panels and batteries require high-cost replacements. A temperature difference is almost always available because most often pipelines carry fluids that are hot. We sell our Power Pucks to sensor manufacturers; resellers have service and maintenance contracts with the manufacturers' customers. A key is for our products to be IS certified, a significant milestone we achieved earlier this year. The major risks are the time cycle of adoption from pilot to full deployment, and the time required to build strong relationships with market leaders that are typically very large multinational companies."

On the wearables side, the available temperature differences are extremely small, and it is where the R&D funded by the National Science Foundation (NSF) Small Business Innovation Research (SBIR) program is important. The company is developing enabling, modular technology, an embedded renewable power source that can be adapted to most wearable applications. The TEGwear product development roadmap is longer and involves integrating Perpetua's component into other products such as wrist and strap products, apparel-based accessories, and sensors. It would be akin to the slogan 'Intel Inside' – in this case it would be 'Powered by Perpetua'. The value proposition for wearables is that body heat extends the life of the power source between recharges. Depending on the power levels required, recharging cycles may be extended from daily to weekly, monthly, or even never.

This business model necessitates partnering with product development teams of established companies. It would involve aspects of product integration; thin-film-based small form factors; flexible electronics; lightweight, reliable packaging; and high-volume manufacturing. The TEGwear program has good traction in the sports and fitness market segment. Adjacent segments are location monitoring, medical, and crossovers into location monitoring for healthcare for say Alzheimer's patients. As with any consumer product, 'wearables' is a dynamic market offering multiple opportunities. It reminds Nick of what Dave Packard once said to him - 'an embarrassment of riches, a company can die of indigestion rather than of starvation.' The consumer market is often whimsical and ephemeral, but the right commercial partner selections can help ameliorate this risk.

Do you plan to manufacture in the U.S.?

"Yes. We manufacture Power Pucks in the U.S. and we plan to continue to manufacture Power Pucks and our other industrial products here in America. On the wearables side, commercial volume increments will be millions - in fact, a million may be a relatively small run. In this market segment, we use a component-supplier model and, because of the large volumes, we are assessing foundries around the world. There are significant benefits to partnering with high-volume manufacturers as strategic partners for wearables."

Financing

Startup funds to develop the technology consisted of angel money, a seed investment from the Battelle Memorial Institute, and grant funding from the Oregon Nanoscience and Microtechnologies Institute (ONAMI), the University of Oregon, Portland State University, and the Department of Homeland Security. Nick, an angel investor, was at that time scouting for promising new technologies in various sectors such as energy, water, and healthcare – technologies with one degree-of-separation from the mainstream but with mainstream potential. Thermoelectric technology offered this appeal. Perpetua's Series A investment was a mix of angels, a small VC firm, and a strategic investor from the manufacturing sector. Several tranches of funding followed as the company hit technology and business milestones. Company management also consistently participated in the financings. In 2010, Perpetua obtained an NSF SBIR Phase I award ($106K), a Phase II award ($500K) in 2011, and a P2B award ($500K) in 2012.

"The NSF awards not only funded critical R&D but are a badge of honor; it offers cache and credibility and helps open doors. We found much broader topic areas to apply to; with other SBIR programs internal requirements drive topic selection. NSF has a strong reputation among tech-savvy investors who are aware that its funding is based on several rounds of technical due diligence performed by experts from around the country. The process of preparing and submitting a proposal is not easy and requires months of arduous work. I distinctly recall many late nights. But in the end it's all worth it."

Other important considerations are time-to-market; a judicious combination of federal, private equity funds, and strategic partners; and timing of venture capital infusion versus stage of technology development. There is a need to find and work together with the right business, financial and industrial partners, to coach the research team to help understand market needs and drivers.

"There's a definite learning curve to working with the federal government; some states, including Oregon, have programs to help. Another consideration is that SBIR funds can only be used for R&D, not for business and market development."

Perpetua is financed with both Series A and B capital raises of $6M that are in-line with national averages. To this is added a mix of grant funding, non-recurring engineering income from development contracts with commercial partners, and revenue streams from activities involving both industrial products and wearables –

"We used Series A funds to get the technology out of the lab environment to a marketable product, and Series B funds for customer engagement and building partner relationships. To develop hardware all the way from the lab to full systems integration in the field is a long, many-years slog. It's a grave issue in America that we are at risk of losing the art and skill of creating cutting-edge hardware products. Hardware startups must be very creative to raise funds, tapping into state and federal programs, universities, NRE, VC/Angels, strategic partners, and customers."

No single entity has a controlling interest in Perpetua. Seed and Series A investors have continued to participate, and stay onboard through each subsequent funding event, even as more investors are added. The Board of Directors consists of Jon, Nick, and three representatives one each from the Series A and B rounds, and one from industry.

Company Growth

The key to success so far has been the reputation and experience base of the initial group, the HP tie-ins, and other strategic partners. Perpetua has been a great beneficiary of HP's local presence that has attracted world-class talent to Oregon. Paul, Marshall, and Nick were all at one-time employees of HP.

"In some cases, it can be hard for folks from large companies to join a startup, but HP has a long history of invention and a culture of being scrappy and innovative. There's nobody I've met like Nick who has the capability to comfortably speak at a strategic level to a Fortune 500 CEO and at the same time get down to operations at a small startup. Nick has been key across all functional areas. Ingo Stark, Kevin Thompson, and Marcus Ward have been the key players in our technical developments. Andy Zaremba, Darren Podrabsky, Janet Dobbs, and Jerry Wiant are important for sales and marketing in our two market segments. We try to architect the pairing of experienced industry professionals with those newer in their career path to provide a bi-modal mix of mentor and mentee. For example, Nick is my mentor, and Paul and Marshall have been mentors for Marcus. This combination has allowed continuity for company personnel, and makes it exciting for team members at all stages of their careers."

How do you judge people?

"The business of people requires immense experience; it's a tough question to answer. In a large company like HP, there's plenty of support from the human resources department, and access to recruiting tools. For startups, the best tool is leveraging existing resources – the current team, and the existing relationships and networks. Much of our recruiting involves word-of-mouth, first- and second- degrees-of-separation. The biggest mistake we try to avoid is hiring opportunistically rather than strategically. We go through a disciplined process of assessing and then matching needs. We generally evaluate along two axes – skill and will. We rely on our network to find candidates who qualify for the 'skill' part. On the 'will' side, we must recognize that not everyone has the disposition to be startup-friendly. There are basic byproducts of startup frugality – not flushed with cash, we typically offset a reduced salary and benefits package with an equity upside. This by itself filters out a few prospects. There's also a nice snowball effect with a few great people joining the company early on. At the start, on the technical side, we needed an expert in semiconductor deposition and manufacturing – Paul fit the bill. He also had access to President- and VP-level folks at Intel and HP. When he describes to them our needs, they readily provide him a shortlist of candidates. We have been fortunate in building a solid team from the beginning."

Can you describe the culture at Perpetua?

Nick - "The founders were Jon and a few others from HP, a company historically known for its culture of collaboration and innovation. We follow these precepts in all things we do. The qualities espoused by a leader shapes culture. There are lots of jokes around our frugality. We are an egalitarian organization – no job is too big or small for anyone. We don't have placards and slogans. I told Jon in our early interactions that this is his book to write, I am only a footnote."

Jon – "I believe in a leadership team that is characterized by 'humble confidence' – we want people with top skills and high will, but we also want people that balance confidence with humility. I think Perpetua interacts with industry in a very positive way, with quiet confidence. Some companies in our industry follow a different culture and come across as aggressive and arrogant. When we interface with customers, our style is much softer. It is relatively easy to get people excited with Perpetua's products – they are well-designed, have interesting form factors, incorporate renewable energy, are used in dynamic market applications, and are quite frankly 'fun'. We don't have to be pushy or aggressive, and there doesn't need to be a race for a hasty sale or a quick hit. When we focus on creating value first, the financial rewards naturally follow."

Key Relationships

Both Jon and Nick believe that NSF SBIR funding has played a valuable role. Funding from this program was demonstrable evidence to other players in the ecosystem that Perpetua's technology and its development roadmap, after diligence from both technical and business experts, is a worthwhile pursuit. It has offered credibility and a signal that others believe in the company's vision.

"We leveraged our NSF SBIR relationship by tapping into the broad set of benefits the program offered, such as assistance with preparing the Phase II commercialization plan, help provided from manufacturing consultants, and the community college supplemental funding. These are structured, in our case, to help transition the technology development of 'wearables' to a set of marketable products. The NSF SBIR program dovetailed perfectly for us. Phase I helped us develop our first working prototype – it makes a significant difference when we have a demonstration model to prove that useful power can be generated from the human body. Midway through Phase II we started to work with a commercial partner, and then the Phase IIB supplement further cemented our relationships with multiple industry partners. If it wasn't for the NSF SBIR program this technology development roadmap would not have been possible. We also collaborated with ONAMI to overcome a few technical hurdles. One other NSF benefit was that it allowed us to participate in Eureka Park at the Consumer Electronics Show. It was a huge deal – it helped us target potential customers and partners, provided a chance for direct conversations, several of which have blossomed."

The Startup Experience

Jon and Nick are optimistic by nature and believe Perpetua will continue to succeed, that it is only a question of how big and when. When asked about serendipity and luck, Nick said - "These are always at play and not uncontrollable, but chance favors the prepared mind. Luck plays a role but looks kindlier upon those who are prepared. Many chance encounters have led to positive outcomes although one never knows when it happens. As you look back, a seemingly innocuous conversation in the past can have a major impact. For example, I was on a Governor-appointed committee along with a development leader from an Indian tribe among others. We got to talking and this chance encounter later led the tribe to participate as a major investor in our Series B round!"

What about the ups and downs of a startup?

Jon - "We've had several of these - sometimes they both even fall into the same category! As a founder and company leader, I feel deeply responsible for the team. It is toughest when the bank account is slim – worrying about payroll is gut wrenching for anyone, and even more so when you've hand-selected your teammates. Fortunately, even during the toughest times, through the recent recession, our team has risen to the occasion and used flexible means of getting us through. This has only further strengthened our bonds. In terms of ups, we've had a lot of 'high-five' moments when we have had significant customer wins, particularly with first sales to Fortune 50 companies. I feel very fortunate that we've had enough of those to keep us excited and energized."

Nick - "Business situations can be handled one way or the other, but for me the most important factors involve relationships – to create, in a startup, a family environment, a band of brothers and sisters. As leaders we inevitably face tough situations. The best part is the blossoming of folks, the shared wins from a relationship point of view rather than merely accounting for dollars and cents."

Regarding personal qualities, Nick finds Jon never in a role where he does not lead by example; colleagues respect that in Jon and are eager to follow his leadership to a place they are compelled to seek. Jon thinks Nick is a practitioner of a subtle, soft style of leadership that simply comes naturally –

"Nick has been very successful to the point that he doesn't have to do any of this financially, but he is still highly motivated to give back, to help others. I've learned more about business in the last five years from Nick than in all my other thirty-six years of life. His time, energy, mentorship, and skills transfer have been such a blessing for me."

" I think we both share a similar work ethic and are driven towards things we believe will provide a true sense of accomplishment – the tougher the challenge, the sweeter the reward. I feel very lucky to have Nick in my life."

Jon, what is your advice to an entrepreneur?

"I think it's really important to choose a business and product that you love. Doing a startup tends to be all-consuming and doing something you like can carry you through the tough times and keep you motivated. I think it's also important to be comfortable with responsibility and uncertainty. Be prepared and excited to be a 'jack-of-all-trades' - to be a business development professional in the morning, a product development person in the afternoon, and a money-raiser at all times. Try to surround yourself with great people with complementary and different skill sets and experiences than yours. Choose partners that you truly care for beyond the business. If you choose to be in the leadership role, you will be truly responsible for the financial wellbeing of your coworkers."

"I think to be a good startup leader it helps to be a glass half-full optimist. Every day could be the worst or the best – make it the best. Patience is also a positive trait for startup leadership. The process takes time, is a lot of hard work and involves sweat, inspiration, and persistence. The highs are as good as it gets. I've really enjoyed our journey at Perpetua and wouldn't have it any other way."

Where was this startup at the end of federal government funding?

Perpetua raised $4M private equity financing from Baldis Holding Company LLC. The company focused its commercialization efforts and resources in the following three areas:

1. Continue to grow industry awareness of its energy harvesting technology by actively participating in industry trade shows and conferences.
2. Expand its customer development agreement efforts.
3. Spin-off its thin-film thermoelectric energy harvesting technology into a newly formed company, Thermogen Technologies, Inc., allowing Thermogen to focus on marketing the technology to the "Wearables" and "Internet of Things" markets.

Perpetua's spin-off, Thermogen, is utilizing prototype thin-film thermoelectric modules manufactured in its laboratory to fulfill its engineering contracts with customers. Today, there are two different sized prototype modules for integration into wearable applications. In 2014, Perpetua drafted and filed a United States patent application (No. 14/267,802) for a thermoelectric generator module for a wearable thermoelectric generator assembly. Revenues total approximately $2.8M. The number of employees in January 2010 was eleven. This number has now grown to 16.

The technology Perpetua developed is well-suited for low-power wearable electronic applications including sports and fitness, health and medical, and location awareness devices. Today, Thermogen markets energy harvesting components to OEMs of these types of devices at various tradeshows and conferences, such as International CES and World Mobile Congress. Thermogen is currently engaged in engineering development contracts with select companies in the wearable electronics market segment.

Thermogen will market the wearable thermoelectric energy harvesting technology under the registered trademark of TEGwear®. It currently utilizes its own direct sales team for obtaining joint development agreements. Direct sales are anticipated to continue long-term for technology integration with select Internet of Things (IoT) market leaders. Joint development agreements provide non-recurring engineering (NRE) revenue opportunities with customers in all IoT application areas as Thermogen works to build up its high-volume manufacturing capabilities.
Thermogen has been able to benefit from industry partnerships and collaborations to augment a lean marketing program that has yielded cooperative demonstration displays at various technology trade shows worldwide to expand industry awareness. Thermogen has capitalized on

co-marketing opportunities with Texas Instruments (TI) and STMicroelectronics (STM). Both highly respected technology partners manufacture complementary energy harvesting related system components that OEMs desire in their solutions.

Today, Thermogen is leveraging prototype modules produced in its lab to educate the market about its technology and attract product development engineering engagements that will lead to high-volume manufacturing contracts. The focus ahead will be to further educate the market on how to successfully integrate Thermogen's technology to maximize its energy harvesting capabilities and to build out volume manufacturing capabilities. Thermogen's leadership is actively engaged in numerous strategic partnership and venture capital funding discussions.

Where's the startup now?

Perpetua exists so that customers can achieve the benefits of long-life power. Using energy harvesting technologies, Perpetua products function like renewable batteries Customers can avoid battery replacements and line-power wiring costs, increase usage rates, and expand into a wider range of environments and applications. Perpetua's plug-and-play Power Pucks® power the world's leading wireless sensors and are available for purchase from Perpetua, as well as directly from Emerson and GE. Applications range from powering pressure, temperature, vibration, and other sensors within industries spanning metallurgy, chemical processing, power plants, oil and gas, and many others. Perpetua is continually innovating and thrives on bringing the benefits of long-life power to the market.

Perpetua, provider of renewable energy solutions for wireless sensors, has announced the acquisition of Energy Harvesting Technologies, Inc., d.b.a. Thermo Life® Energy Corporation ("Thermo Life"), a leading developer of advanced thermoelectric technology. Thermo Life first introduced its 3-volt thin-film thermoelectric device in 2001, under joint development with Diamond Tooling System, GmbH. Since then, Thermo Life has obtained multiple patents for its unique thermoelectric processes, representing over a decade of research and development efforts. Thermo Life's technology and solutions will be integrated into Perpetua's existing product line, enabling miniaturized product solutions for use in powering wireless body area networks and other key applications. Ingo Stark, a world-renowned expert in thermos-electrics and CTO of Thermo Life, joins Perpetua as Chief Scientist and will head thermoelectric research.

"With the addition of Thermo Life, Perpetua offers one of the most comprehensive thermoelectric energy harvesting platforms in the industry. We can now offer thermoelectric solutions for powering body-worn sensors all the way to industrial monitoring and process control systems," said Jon Hofmeister, Perpetua President. Additionally, Dr. Ingo Stark will enhance the thermoelectric science community already internationally respected here in the Pacific Northwest.

"We are very excited to be joining forces and bringing the value of our combined assets to a larger audience," said Ingo Stark. "We share Perpetua's vision of offering full product solutions to the wireless sensor market and making thermoelectric solutions easy to understand and use."

The Perpetua thermoelectric energy harvesters were quite recently demonstrated at the Exelon Innovation Exposition held in Washington DC in June 2017. The exposition, which included key Exelon operating companies and personnel, showcased Perpetua energy harvesters, and demonstrated Perpetua products powering wireless instrumentation. Many of Exelon's generating facilities utilize extensive networks of steam pipes which provide widespread heat sources ideal for Perpetua's Power Puck and Power Tile thermoelectric energy harvesters. By using the energy harvesters to produce power for Exelon's process instrumentation, the company can reduce or eliminate battery replacements for wireless instrumentation, decrease installation time and reduce the overall costs of wired devices. Exelon is a Fortune 100 company that works in every stage of the energy business: power generation, competitive energy sales, transmission and delivery. Exelon Generation is one of the largest U.S. power generators and has approximately 32,700 megawatts of nuclear, gas, wind, solar and hydroelectric generating capacity comprising one of the nation's cleanest, lowest-cost power generation fleets. The company does business in 48 states, D.C., and Canada, and had 2016 revenues of $31.4B. It employs approximately 34,000 people nationwide.

Case Study Questions

1. Discussion: A startup operating in two different markets – industrial and consumer
2. Team building using a mix of experience and young talent
3. Analyze this startup creation process
4. IP strategy
5. Business strategy for multiple business models
6. NSF SBIR Program forces consideration of 'wearables' market?
7. Start-up valuation exercise
8. Risk elements and how to manage/mitigate them
9. Growth and exit strategy
10. The need for certified products
11. Partnering with large multinational companies
12. A matter of humility
13. Take-away(s)

9   *Let Us Not Allow Milk to Spoil*

Technology Sector: Materials
Startup: Promethean Power Systems
Website: www.promethean-power.com
Location: Cambridge, Massachusetts

Federal Funding Timeframe: July 2011 – September 2015
Funding Amount: $0.76M

This narrative is a story of: The need for private equity funds before federal funding; bootstrap growth; technology pivot(s); business model changes; scale-up/ production challenges; foreign subsidiary established; operations in multiple-countries; significant societal impact

Startup Formation

Sorin Grama grew up in communist Romania where private enterprise was not allowed, and the notion of starting a business of any kind was entirely foreign. His family immigrated to America when he was 18. He was always independent-minded - in college he did not join many clubs or fraternities like other classmates and instead preferred to take up internship opportunities at computer and computer networking companies. After graduation, he worked for National Instruments. He enjoyed it and was for a while a technical sales representative where it seemed like he had his own business, operating independently with his own schedule. He later joined a small Bay Area consulting company involved in hardware-software integration projects. He found this experience more entrepreneurial working with two other business partners. After a few years of professional work, Sorin decided to pursue a graduate degree at MIT in Engineering and Management. In graduate school, he quickly gravitated towards entrepreneurial and product development-type courses through which he gained important lessons and insights, and an appreciation for the product development cycle. In the campus, he participated in entrepreneurial activities and business plan competitions. In one such competition Sorin helped write the business plan for a team of Ph.D. students focused on solar concentrator cooling technology. On their second try, they won the second place $10K prize.

Members of the team that won the business plan competition were soon recruited by companies. Sorin did not want to work for a big company and was hoping that the winning business plan could be turned into a startup. Many people advised him on India, he read many reports on the Indian solar market, but he had never been there. With the $10K prize money he decided to travel to India to study the solar market there - the solar concentrator was after all designed for a developing market. Sam White, who was working for a company in Boston and is now Sorin's business partner, went along. Sam was also part of the team that had won the business plan competition - anybody can participate if at least one team member was an MIT student. The two were of the same age, had professional work experience, and quickly became friends. They spent a month travelling in India trying to understand the market and its opportunities. This was a classic case of a technology looking for a market – using the sun to produce hot water and electricity but for what market?

"This is not the best way to start a business. It should be the other way around - find a market need and design a technology solution for it; we had it upside down! Lots of people in India said this is great, we can use it. People were openly curious about the two of us from America – it was easy to schedule meetings with many high-level players. Sam is great at cold-calling folks that we were introduced to through MIT – he's a people person, he's OK with a person slamming a door in his face. We kept looking, talked to lots of business and industry people. In one case, we had the owner of large sugar mills send a car to show us around, we were treated like royalty; in other places, we roughed it out following our own plans. One Saturday with time on our hands Sam arranged an appointment with the Managing Director of Bangalore Dairy - he had a team of engineers waiting for us. They were not too interested in the solar concentrator. Instead, he told us about his problem – milk spoilage. After that chance encounter, we met people at two other milk dairies and heard the same message. We decided to explore this market – India is the largest milk producer in the world; the market is sophisticated, well-organized with clear and lengthy supply-chains."

Technology and Initial Financing

The other members of the team that won second place in the business plan competition decided to continue their studies at MIT and focus on the African market through a non-profit organization. Since the solar concentrator technology was not well suited for the new market opportunity that Sorin and Sam had discovered, the two partners decided to start with a fresh, clean slate[15]. Now the duo had but an idea, some market knowledge gained, and contacts made from the trip but no technology. They decided to conduct a thorough product analysis for this market for six months, perhaps a year. They were unmarried and did not have other commitments. Sam had quit his job before going to India.

"When you finish school is the best time to pursue startup leads.

We met a social impact investor who made bets on the solar market in India and Africa. Based on the earlier solar concentrator technology and our limited knowledge of the Indian market, we asked him $25K but he said take $50K, go figure the opportunity some more. He found solar and milk spoilage an interesting market segment that aligned with his goals. We incorporated Promethean Power Systems ("Promethean") at the end of 2007 to solve this dairy industry problem, to figure out how to make a solar power contraption to chill milk. We researched and analyzed the many ways to cool, narrowing our efforts down to an approach that will work. We first went with thermoelectric technology using Peltier junctions, an elegant way, but it was not very efficient and not well-suited."

Sorin spent nine low-budget months analyzing and evaluating. He built a working prototype using solar and thermoelectric technologies in his apartment, using parts bought from Goodwill and eBay, and equipment made available to him from National Instruments, his previous employer. He went to a thermoelectric expert to build a larger unit, but the system turned out to be inefficient taking too much time to cool. An Australian company over-promised and under-delivered – not too familiar with the technology Sorin lost precious time and money. Meanwhile Sam was talking to people, trying to better understand and assess the market. Both were devoting 100% of their time to the company but doing part-time work to pay the bills. On the side, Sam was selling solar installations for a local installer while Sorin was writing and selling solar market research reports.

"This was real entrepreneurship; I look back with great nostalgia, it was our best fun time; people gave us equipment they didn't want, and we continued investigating. There was pressure to produce from the investor but the only pressure we felt was an internal desire to get something done, the best kind of pressure. An article about us in MIT Technology Review and press coverage of solar power in India got the attention of another investor. Sam cold-called him after having learned of his interest in solar power - just like that the man said – how much of your company do I get for $1M? We didn't know what to do. When we asked him for his references, he went silent, and we lost contact. We were bummed and dejected – we didn't know how to close a deal with such people; then after a few weeks, he came back – I'll give you $500K, go get the other $500K from someone else. We couldn't find anyone, but he gave us $500K anyway. We signed the investment papers right before the market crashed in September 2008 and Lehman Brothers went under. If we had waited another week, we probably wouldn't have gotten the investment."

Promethean now was a real company with stock certificates, formal lawyers, and cap tables. The work on solar and thermoelectric paid off in a way by allowing the company to obtain this half-a-million-dollar investment. They could hire people, talk to partners, it lent credibility and people began to pay attention. They began to investigate commercially available vapor compression technology while at the same time not wanting to forsake solar power. In early 2009, Sorin and Sam went back to India for a month to conduct a more thorough market analysis – a classic market study following a drop of milk from cow to consumer – to understand the entire supply-chain, the weaknesses and holes in the cold-chain, the user, the buyer, the milking nodes, the rush to collection points to avoid spoilage, and then to the dairy plant. A milk chiller in India needs a generator because the supply of electricity could often not be depended upon. This is where the two partners could apply an innovative technology to mitigate or solve the milk spoilage problem.

"We gained good insights from the trip and began to better understand the dairy industry in India, learned to speak the language. I am more of a hands-on guy, I talked to the farmers, watched them milk cows, and went in auto rickshaws crammed with milk jugs to the dairy plant. It's fun now to recall all that we went through. If there's a lesson learned here, it is that to thoroughly understand the problem, first look at what's being done. We came back and engaged an engineering design consulting firm in Boston to help design a product we had in mind."

In 2011 Promethean delivered its second-generation solar-powered refrigeration system, still a research prototype, to chill 500 liters of milk per day, to India's largest private dairy owner. This customer had encouraged Promethean, saying he would purchase multiple systems if the price-performance met his expectations. It was built in Boston and shipped to Chennai, India in February 2011. The installation was an important event for the dairy and included a gala unveiling amid high expectations. The managing director arrived with great show and pomp – upon inspection, he was immediately disappointed, said it was too expensive, took up too much space, was not cooling milk fast enough, and it was impossible to have such a large, complicated unit in dozens of villages and towns. Promethean had spent nine months building this one prototype and had depleted their entire funding – it was a crisis! After much deliberation, the team now including a few Indian employees wondered if they would have to cast aside solar power.

"Everything we worked for was to use solar power, but it was turning out too expensive, the unit was difficult to power up, moving it to village locations was logistically hard, milk chilling capacity was low, and installing solar panels on roofs was too complicated. I was very reluctant, all my hard work and the expertise I had developed in solar power would go to waste, investors had invested in the solar idea - it was part of our fabric. I went off alone for a few days to agonize and think about the vision and principle to use renewable energy versus the practical market problem. All along, the customer remained positive, didn't want us to give up, and wanted us to work to make it smaller and cheaper."

The team finally decided to give up on their dream of using solar power and focus instead on providing a workable solution to the real problem at hand. They resolved to build another prototype – it was much more cost-effective to do this in India. They discarded everything except the chiller. This smaller prototype also did not perform well enough. The same problem persisted – how to chill milk when there is no electricity supply? Suddenly it dawned on them that the energy storage device was the key. When they had previously relied on solar power, a large tank of chilly water was used as an energy storage mechanism for night-time operation or as a backup during cloudy days. The critical insight was that the energy storage device was the key element which can operate together with grid electricity like a battery to fill the gaps in grid power. Their thinking was obscured by the need to use solar power and they had missed the solution, the true innovation.

Fortunately, in June 2011 the company was awarded a Small Business Innovation Research (SBIR) Phase I grant from the National Science Foundation (NSF). The timing was uncanny – if it had not been for this grant, they would surely have run out of money. During the crisis, Promethean had to lay off the two fulltime engineers in Boston, and so had to rebuild a team to work on the SBIR grant. Sorin and Sam went back to Boston, analyzing, experimenting, and iterating towards a more compact, less expensive, and more efficient energy storage device. The drivers were the choice of material, battery construction, material packaged to be super-cold at one end, and then be able to quickly chill 500 liters of milk without electrical power. In the meantime, Promethean also obtained an order for five units from a few Indian dairies that saw much improvement in the new system, including the dairy that had rejected the earlier system.

Customer Discovery and Business Model

Sam is responsible for business development, spreading the word about Promethean through cold calls to potential customers and investors, online video clips, press releases, and attending conferences.

"We go to the annual dairy conference to meet potential customers there. In fact, we met our biggest customer at one of these events. We never display a product. Instead, we make presentations talking about quality, operating costs, cold-chain market trends, and the technology behind our product. An American firm working in rural India attracts attention. Sam is easy-going, friendly, likes to interact with people – we need this in a startup, a combination of engineering and marketing. As for the business model – it's a straightforward equipment sale. It's that simple. A milk dairy buys it as part of their capital expenditure. Large dairies purchase lots of equipment, they have the money and manpower to train support personnel. The farmers benefit because they have a market for their milk, the milk doesn't spoil, and they don't get cheated by middlemen. We sold five prototypes – we hired a design firm to make them look classy.

We had reduced system size by half and doubled the capacity to 1,000 liters a day. The fancy demo system worked but was still expensive. It required a third design iteration for a commercially viable product."

One early system in a village outside Salem in south India has been in operation for almost three years. In 2012, Promethean started installing their first systems in five big dairies. They outsource the fabrication of major components, but design, integration, testing and installation are performed in-house.

"One customer installed a system but does not use it anymore, another had too many problems with the powerline transformers, and the two others are still running. Some customers put a lot of effort to make it work properly – the customer as partner is a key element. We started to get more orders and so in December 2012 we set up a joint venture with an Indian partner making ice-making equipment. The partner had the required infrastructure, understood the supply-chain, had a good network of suppliers and contacts, and had been in business for twenty years. We co-located in unused space in his factory for a while before moving out to a location in another nearby city. Promethean U.S. is the majority owner of the Indian subsidiary. In January 2013, we got an order for fifty units. We decided to perform another iteration to further reduce the cost, develop a production-ready design that was simpler and easier to manufacture. It was cart before horse – had the order and money in hand but not yet the final design. The NSF SBIR Phase II came through in March 2013. In Boston, we systematically analyzed every single component for need and functionality to simplify the overall design."

Soon an Indian private equity firm invested $500K in the Indian subsidiary. Customers were told that at ten units per month, delivery would start in April 2013 and the entire order fulfilled by December 2013. It did not work out that way! There were innumerable problems with the new design – a change of a single component had cascading effects affecting several other subsystems. A new pump, for example, affected heat transfer in many parts of the system which made it not perform to specifications. The Boston team wrestled with the problem for two months – was it a manufacturing defect? It was 24/7 development – re-design and troubleshooting during the day in America, emails in the evening from the U.S. to test and experiment on the manufacturing floor next day in India. It was finally resolved – the problem was traced back to a tiny mesh filter in one of the lines that was rapidly clogging up leading to reduced flow rates in the piping subsystem.

The first commercial system was at last ready the end of July. The unit was installed and inaugurated in a village with an elaborate ribbon-cutting ceremony accompanied by prayers and rituals. Powerful players from the dairy industry gave fiery speeches. It worked perfectly in the factory but in the field at a village site the chiller unit did not cool down to target temperature, and not quickly enough. Sorin arrived at the village to fix the problem. He noticed a pool of blood, some of it on the machine. The locals had sacrificed a goat to exorcize demons they insisted prevented the machine from working properly. The problem was soon fixed. The villagers rationalized that it was so because of their animal sacrifice. It turned out to be one of Promethean's best systems, designed to collect 400 liters a day, it now collects 1,300 liters, exceeding by large measure the system capacity and necessitating the emptying of milk twice a day.

"The dairy that purchased the system hired an attendant to collect milk from village producers and to ensure proper functioning and cleaning of the system. For his work, they paid him a commission. The entrepreneur in him makes him ask me if he can get a second machine to make more money. I enjoy watching the smiling, happy faces of simple rural folk. The last of the fifty units was delivered in October 2014, almost a year beyond what we had originally planned! We have learned many lessons out of this experience, specifically regarding variability in production processes, testing, and quality control in small-scale Indian manufacturing. The quality of outsourced work is not good – I don't yet have a good handle on training and testing. The NSF Phase II was amazingly timely, just when we were in the thick of things developing the new design in Boston. We have one more design cycle to go through to bring down the cost further. In the first fifty units, we quoted a price that we thought was reasonable but the finished design cost more, so we made no profit on these systems. The investors are OK for now, but we hope economies of scale will lead to larger margins. The systems now are profitable. At last count, we have delivered eighty units, with seventy-five of them active in the field."

Company Growth

Sam and Sorin started as equal partners. Throughout the first year they were the only fulltime employees - Sorin engaged in product development and Sam gaining insights through market research. Upon receipt of seed round funding at the end of 2008 two additional members joined the team. Along with two interns, and an outside contractor, the team consisted of seven people all based in Boston. They set aside stock options that had helped recruit the two early hires. These two team members were unable to exercise these options because they had to be laid off during the vesting period. Today, the two founders are majority shareholders, and the early investors are the minority owners of Promethean USA. The Indian subsidiary is majority-owned by the U.S. entity with Indian investors owning a minority share in the subsidiary.

Are there key personnel in the Indian subsidiary?

"Yes, there is one person we consider key – a disciplined professional from the dairy industry with thirty years of experience in equipment manufacturing processes and organizational skills. He's been with us for less than a year, but he's already made an impact. One concern is that if a key person like him leaves, some of the others may leave too. We plan to give our key employees stock options in the Indian subsidiary."

What do you see as people-challenges in a two-country operation?

"Recruitment is a big issue – so far this has been opportunistic in both countries. If a qualified person seems passionate about what we do, we hire him. Because this is not the usual office kind of work, people find it exciting and an opportunity for a career shift. Some who have joined us this way are very good, but this is probably not an optimal method to build a team. We have not yet recruited for specific skills – our company is more mature now and we need to become more systematic. All startups must manage this transition. We do need a high-caliber engineering manager to become the CEO of the Indian subsidiary. Right now, I am the CEO of both companies but especially in India with its complex social structures, ancient cultural norms, and other operational nuances I am not the best person. It's also not effective to overload one person with too much responsibility. I'd like to focus on product design and engineering and have someone else manage operations and capital-raising."

What are some differences you see operating in these two countries?

"It is more difficult managing engineers in the U.S. – they tend to be more independent and self-minded. Young engineers in America are more opinionated and argue about the path and direction a startup takes. In India, they defer to authority and thus are easier to manage. They prefer to follow orders but are hardworking, thankful they have a decent job, and feel less entitled. This can be good and bad – it's hard to find outside-the-box thinkers; accustomed to rote learning and doing well on exams but not hands-on, not creative. There's too much red tape and documentation to register a business and set up a manufacturing facility in India. There's immense paperwork involved even to ship one piece of equipment from the factory to the site. Another set of paperwork if you have to ship it back for some reason – have to find all kinds of creative ways to ship across state lines. To take anything out of a factory, say a prototype out to the field for trials, not selling just field testing, the documentation can be maddening. Bringing it back after field trails is another set of headaches. The tax system is a nightmare – every state has its own system!"

What's your endpoint?

"I'd love to see the India business self-sustaining, deliver milk chillers customers love, make a profit, no need for a blockbuster, just enough to take good care of our employees and customers. Finally, the

Indian subsidiary could be acquired by a larger company. Sam and I are less clear about the U.S. entity, maybe a steady licensing royalty stream from the Indian company and then either be acquired by a multinational firm or buy out our patent rights. I feel that we need to find a large strategic partner or buyer who can help us overhaul the cold-chain market in India – it's too big a scale. Also, being in the trenches, fighting every day, so buried in detail, it's hard to see the big picture, to cultivate a big vision. The Indian subsidiary needs new blood, someone else to scale it, with a different attitude and skill set. Investors tell me it is rare for founders to be around for seven years. I'd like to pass the company on to good managers, and then explore other research problems and technologies."

Key Relationships

In dealing with large milk dairies in India customer relationships are most important. Some made a conscious business decision to go in the direction provided by the Promethean product.

"We must make sure we deliver our product on time with high quality to make it happen for them. Our customers must be taken care of; and our employees too. Their growth can stall if we don't come through. We have been working with this one customer for five years – from the very beginning they had a dedicated team to work with us on this product, invested so much money and resources into integrating our product in their business operations. We cannot let down such customers – one needs only five such big milk dairies. Together their market share is 15-20% of the biggest milk producing country in the world."

Why not use this technology for fruits and vegetables, and vaccines?

"In fact, we are doing just that – developing a cold storage container using the same energy storage system for other perishable items. Our first customer is a lettuce and tomato grower. They plan to use our product right after harvesting and before shipping to market. It's a big growth area for us. We did look at medical supplies – there are quite a few companies in that market. There are regulations crafted by the World Health Organization (WHO) that we have to abide by, it will require a drastic scale-down of our current product involving lots of technology development effort, and so we decided to continue to focus on milk, fruits and vegetables."

The Startup Experience

Sorin left Romania after high school. His journey embodies the classic American immigrant experience – an outsider with not too many friends he was highly motivated to learn this new culture and study hard to take advantage of the American Dream. He believes that one can more easily transcend tradition and force change when you are not part of a system.

What made you want to be an entrepreneur?

"I often wonder about this. In Romania, private enterprise was not allowed, and I didn't grow up with an entrepreneurial spirit. I guess it is partly being an immigrant, not being part of the system – I was mostly on my own. When I worked at a small systems integration company, I had a lot of freedom and independence, and I liked that. This startup thing just happened – I was not inspired by a role model, and no one talked me into it. A kind of stubbornness was part of it – people not very encouraging asking me why I was doing this, wasting my career. I also liked the challenge of it."

Is there an element in you of wanting to make a difference?

"This also just happened, it was not by design; I see a problem, I like to fix it – it's the engineer in me, always motivated to find a better solution. I get a lot of satisfaction from seeing people using our product, seeing their happiness. It feels good to make a difference in people's lives. You asked about a high point: My first job was with National Instruments in Austin. I worked there for five years, three in sales, and cultivated a set of good relationships while there. Even now we use their equipment and instruments. I maintained those connections and last year they asked me to deliver a keynote presentation to three thousand people on how we use their products. Friends and past co-workers told me they were amazed how far I had gone. It was greatly satisfying to round out the circle in my mind – to be back where long ago I had started. After so many dark days of challenge and struggle, I cherish these moments because they don't come often. And a low-point was when an important first customer was disappointed by our first solar-based system. For almost a year the fiasco with the first manufacturing business partner was a low-point too – too many low points! It happens to all entrepreneurs, and one should not get too affected by them, after a while you get used to it. A mediocre exit is OK with me, what's important is changing an industry, a whole industry adopting this technology, to see many companies making rapid chillers, not all ours, installed in lots of villages, changing the attitude, all this can be a lifetime's worth of work."

Did luck play a role in your journey? And what of your Indian experience?

"Yes, some element of it – lucky to get our first investment, NSF funding when it came was fortunate and timely. You make your own luck.

No amount of planning in the world can make a company successful. There were many chance encounters – our first customer is a good example. My advice to a young person starting out is to first ask about motivation. If it is to be famous, make money, it's the wrong way. Instead, if it is – here's a problem and I'd like to fix it, and this is how I am going to fix it. What's his approach - is he passionate about doing something useful; fame and wealth are byproducts. Being originally from Romania, being in India is easier to deal with; less of a shock for me to come to India - India reminds me of my childhood in Romania, the lack of power, the bad roads, the bad infrastructure, the poverty. To me, some parts of India are just like Romania. The immigrant experience in America also helped for I am also an immigrant in India!"

Where was this startup at the end of federal government funding?

Promethean designs and manufactures refrigeration systems for cold storage applications in off-grid and partially electrified areas of developing countries. The company's products enable food suppliers to reliably store and preserve perishable food items such as milk, fruits, and vegetables, without the need for expensive diesel-powered generators. This creates a cost- effective solution for cold-chain food distribution in emerging markets, a great business opportunity that could also deliver enormous social and environmental benefits.

The technical innovation is a thermal battery used in refrigeration applications such as cold storage of perishable foods: fruits, vegetables, or milk. Promethean's first commercial application is a milk chilling system for the Indian market where its thermal battery is used as an energy storage device to assist the unreliable grid infrastructure in rural areas. Billions of dollars of perishable foods are wasted annually in India because of inadequate cold-chain supply networks. A major obstacle in setting up a cold-chain network is the lack of reliable grid electricity to run refrigeration systems in villages and farming areas.

To date, the company has delivered over a hundred commercial refrigeration systems with a total energy storage capacity of over 2.5 megawatt-hours. Promethean's milk systems are installed at a village collection center where twenty-thirty farming families deliver their milk twice a day. Farmers receive a good income from their milk which is essential to Indian diet and the rural economy. In rural India, milk generates the cash that pays for daily expenses and the education of children.

Promethean's next commercial applications are in the United States where the battery can serve as an energy efficiency device to shift large refrigeration loads from peak time to off-peak hours. Its battery is currently being used to provide fast, efficient cooling power for a micro-brewery in Boston. Although Promethean's first commercialization efforts are in India where thermal batteries are needed the most, these efforts benefit the U.S. economy by opening a large and growing market to U.S. products and strengthening America's leadership role in spreading innovation throughout the world.

Locally, In India, these thermal batteries help reduce India's dependence on foreign oil (by eliminating the use of diesel generators) and improves the quality of milk and fresh produce for Indian consumers. As a partner and ally of the United States in this region, India has welcomed America's assistance in solving its difficult grid infrastructure problems.

Where's the startup now?

The motto of the company is "Preserving perishables in some of the world's most challenging conditions". Promethean designs and manufactures refrigeration systems for cold-storage and milk chilling applications in off-grid and partially electrified areas of developing countries. The goal is to enable village-level chilling with a range of products that deliver high-quality produce to its customers. What is most exciting is that not only is this a great market need but also a win-win-win opportunity: a way to make a positive impact on quality and costs with dairy processors, better quality products for consumers, and increased opportunities for farmers in villages across India.

From individual farmers to billion-dollar food and dairy conglomerates, Promethean's thermal-energy-based technology makes a measurable difference wherever preserving perishables is necessary. Since 2013, leading dairies have installed Promethean's systems across hundreds of village-level collection centers. This has helped customers to chill milk successfully, without using a single drop of diesel through the company's Rapid and Conventional Chilling solutions running successfully across India, Bangladesh, and Sri Lanka. The impact of these products also percolates to individual farmers, increasing their economic opportunity.

At the heart of Promethean's solution is the Promethean Thermal Storage System (TSS), a patented technology which enables all of Promethean's refrigeration products. The TSS can store and release large amounts of thermal energy and can be applied to cooling applications as varied as comfort cooling or fermentation control. It provides backup cooling power for areas with unreliable grid power, instant cooling power for rapid cooling of fruits, vegetables, milk, and other perishable food products, and load shifting – from daytime to night-time to reduce energy bills and increase energy efficiency in refrigeration applications.

Case Study Questions

1. Comment on applying American technology in a developing country
2. How best to scale in a market that is the largest in the world in milk production?
3. Entrepreneurial nimbleness and the art of pivoting, a much-required skill?
4. The challenges of operating in two entirely distinct cultures
5. Discuss the use of appropriate technology to solve real and pressing problems
6. How best to exploit opportunities in adjacent markets?
7. Team building/dynamics – finding the right people in multiple countries
8. The advantage of 24-hour operations especially for troubleshooting product design issues
9. Would additional equity funding be required and why?
10. Identify risk elements and how to manage/mitigate them?
11. Prepare a growth and exit strategy
12. Take-away(s)

10   Helping Wheelchairs Navigate

Technology Sector:	Robotics
Startup:	Love Park Robotics, LLC
Website:	http://loveparkrobotics.com
Location:	Philadelphia, Pennsylvania
Federal Funding Timeframe:	January 2012 – December 2016
Funding Amount:	$0.93M
This narrative is a story of:	The need for private equity funds beyond federal funding; sales/bootstrap growth; technology pivot(s); business model changes; significant societal impact

Startup Formation

Tom Panzarella is the sole founder of Love Park Robotics ("Love Park"). This is his second startup, the first being Freedom Sciences which was acquired by a larger company - "I left Wall Street in 2003, moved back home to the Philadelphia area where I took a position at a machine-learning VC-backed startup in growth mode. My dad owned a manufacturing company outside Philadelphia. He started his career with General Electric, a company he left in the 1970s when he came across a family-run manufacturing contract firm. He saw an opportunity when the owners wanted to transition out of this business over a five to ten-year timeframe. My dad grew this business from a couple of hundred thousand dollars in annual revenues to annual sales of $17M."

At a trade show Tom's father had watched the crude ways that handicapped people used to access their vehicles. He decided to do something about this. Human assistive technology rooted in robotics was outside his professional sphere and his company's core competency, and so he initiated a corporate-sponsored research project, the Automated Transport and Retrieval System (ATRS). He recruited researchers from Carnegie-Mellon University, Lehigh University, and the University of Pittsburg.

"One day my father called me to ask if I'd be interested in ATRS. He said he had researchers working on it but that he wasn't sure if his money was being efficiently utilized. I agreed to help. In six months, we had a working prototype that we demonstrated at a mobility industry trade show. The interest and excitement shown by market players was proof of a successful concept. My dad suggested I further develop this through a spinout from his manufacturing company - thus was Freedom Sciences born."

John Spletzer, after obtaining his Ph.D. from the University of Pennsylvania, joined the faculty of Lehigh University. He was part of the team that Tom's father had assembled to develop ATRS. Tom and John have worked together for over ten years - when Tom was involved in Freedom Sciences and now with Love Park. Tom was the Chief Technology Officer (CTO) at Freedom Sciences and John, who remained a fulltime faculty member at Lehigh, the Chief Scientific Officer. Between 2005 and 2010 they grew that company to $7.5M in sales in the U.S. and Europe. A private equity (PE) firm purchased it in 2010 as part of their rollup strategy of consolidating fragmented small companies in a high-growth sector driven by an aging population and other demographic shifts.

"When I ran into deep technical issues, and didn't have sufficient bandwidth to handle it myself, I could always lob it out to John in his lab. My father was a role model, an engineer who was entrepreneurial, a blue-collar CEO. He instilled in me a work ethic of hard labor, said that I should work a few years for a big company, understand business and learn the ropes. At Freedom Sciences, I was a very small minority owner; Dad's manufacturing company was the prime equity holder. I was mostly focused on technical aspects, and the engineering. When the company was bought, the PE firm didn't want a CTO, and I didn't want to join them either, but they did want me involved on the technical side. I remained a contractor to them on the ATRS product. A two-year guaranteed payment for my performance contract provided money to start Love Park. Initially, I did consultant work but what I really sought was a product company. It would be mobility industry related, directly benefit people, and should extend to outside of the wheelchair space."

Over the years, Tom had carefully cultivated many professional relationships and had a good understanding of the market, a complicated market with insurers, dealers, payers, hospitals, and users. He ran his ideas by trusted people, ideas involving high-technical risk product development, not in the wheelhouse of a typical venture capital (VC) firm. Tom therefore had no choice but to reach out to federal programs. He wrote two grant proposals and was fortunate enough obtain a Phase I National Science Foundation (NSF) Small Business Innovation Research (SBIR) award.

Technology

Tom understands that there must be a market side to everything he does – industry issues, market dynamics, commercial constraints, the need to build partnerships, and what it takes to create a product. At the same time, he does not underestimate the importance of incorporating leading-edge concepts in robotics and human assistive technology into his products. Tom believes that being connected to a university is a great asset and considers John and his lab at Lehigh an extension of Love Park Robotics. In fact, the company continues to fund part of John's research to further enhance the technology content of his products, especially in the areas of traction, mobility, and in decisions on the mix of fully autonomous versus human-in-the-loop control. The essence of the technology deployed is cooperative control involving human and machine working together to achieve mobility goals[16]. Driving aids being developed do not replace but supplement and augment. The machine takes over when the required precision is beyond human capability, otherwise the human remains in control. Payers insist that total autonomy is overkill and will not pay for it. In his mind, Tom believes that the technology being developed be not limited to wheelchairs and handicapped people. It could for example be applicable to materials handling in manufacturing or as an assistant to drone flight control. These technologies are enabled by exploiting the latest advances in real-time three-dimensional (3D) computer vision. In many industries, there has been an explosion of low-cost 3D image sensor use to improve operational capability. Presentations by Tom at various forums routinely generate interest from outside the mobility industry on sensor, perception, and control technologies being developed through R&D supported by the NSF SBIR award. These have led to additional intellectual property (IP) owned by Love Park.

Can you describe the technology pieces that underlie your vision?

"The three critical technology pieces are perception that requires both core expertise in 3D object recognition and handling 3D data in real-time; planning; and control. These are in turn supported by two core principles - embeddable technology to obviate the need to create a new vehicle and thus being able to provide current vehicles with superior performance; and a design philosophy around cooperative control that avoids the off/on switch trap meaning human control is injectable whenever and as required. Planning involves incorporating navigational techniques and perspectives, environmental dynamics, modeling and recognition of specific objects like doorways to enable our product, CoPilot, to go from point A to point B; control – once an obstacle is found, how then to navigate around it using low-level vehicle control, reliably and in real time. Regarding a technology roadmap and key milestones, I am not entirely sure because we don't yet know what shape the products will take. The technology pieces we are developing are like Legos in a box – the question is which set of Legos would a commercial partner use and pay money for."

What about IP considerations?

"IP is a funny thing; patents are expensive. There exists a fine line between core competency (trade secrets) and what should be protected (patents). Patents are often not the most technically interesting but more for market positioning. We have a mix of both, with four patents pending from our NSF SBIR work. Ultimately it comes down to our team and real people using our product in real situations. I try to carryover lessons learned from my time at Freedom Sciences to avoid mistakes in my new startup."

Customer Discovery and Business Model

Tom has met with potential end-users and has thus been made aware of multiple market opportunities. It reminds him about what one venture capitalist said to him – "A startup must be like a heat-seeking missile to sniff out the right market segment."

There are numerous stakeholders involving technology providers, prescribers, regulatory bodies, payers/insurance companies, and users. Occupational therapists, the prescribers, are the most important because they are the ones who evaluate patients, determine the needs of users for their daily functioning, and prescribe machines. On the other side are the payers – what would they pay for, what would they not pay for? For the wheelchair market, a local rehabilitation hospital provides guidance on product direction and features, and important introductions to top-tier manufacturers. A decision to not sell to end-users was made early. Instead, the startup will develop a business-to-business (B2B) enterprise product that they would supply to other manufacturers. Tom knows that direct-to-customer sale is not his forte –

"Many discussions with multiple potential customers have helped us decide not to manufacture vehicles but to supply components to augment wheelchairs for two principal reasons. On the business side, we will be aware of the number of components we sell, and on the technology side, we'd be able to control quality on specific components. Quality is the key – if a component fails then we are done. Using this model allows us to explore the technology space. On the other hand, we may be giving up lots of revenue by not being a device and/or vehicle supplier, but then we would not have to carry the overhead that this would involve. Also, the user community has not much say in the matter. Users don't control the money, that's the problem, it is insurance companies that do."

The company is focused, for now, on the wheelchair market but is beginning to explore other sectors believing diversification to be a good long-term strategy. In the wheelchair market, feedback loops with vehicle manufacturers are long, although prescribers are more responsive. For example, the Good Shepherd Rehabilitation Hospital introduced Tom to the largest wheelchair manufacturer in the world, Pride Mobility ("Pride"), a company with an estimated $350M in annual revenues.

"Our business model involves working with large companies, but they are often slow and bureaucratic. Pride showed great interest in licensing our technology - we have done a ton of work for them over the last two years and were furthest along with this potential partner. They recently purchased another company – now all their time is focused on integrating this new company into their operations. It is frustrating that they have no time for us; our technology complements the products of the company they purchased, a firm that was an after-market provider of accessories to Pride's wheelchair. This other company makes switch-based head arrays that help to steer a vehicle using head movements. That mode of driving is fine with wide open spaces but in a tight space going room-to-room through doors, it's very difficult. We have developed our door navigation system and integrated it with a head array. This is good for open spaces and automatically navigates through doors in indoor spaces. Pride sees it this way - head arrays have been commoditized but nobody they are aware of is doing what we do, so they don't see us as an imminent threat. We could possibly be their next serious engagement. They don't have to move immediately. To them, we are an interesting play, and brand-new. We were days away from a deal with Pride when they acquired this other company – the rug got pulled out from under us. We can't sit around waiting for them to act, and it's important to look at how we can apply our SBIR work to other markets.

"We continue to mature our technology stack. There are opportunities in several adjacent markets. One example from the manufacturing sector involves the use of automated guide vehicles (AGV) for materials handling. This would leverage our core competency in manipulation of 3D data for object recognition in real-time using new lower-cost 3D sensors. The sensors used today in such vehicles are crude and dated. The biggest challenges in the AGV market involve throughput, speed, and efficiency, for example, in moving cases of merchandize from one end of a warehouse to the loading dock. Love Park's localization algorithms can be used to drive the forklift and locate then place the fork into pallet pockets using raw 3D sensor data from cameras to determine what makes up a scene (positions of fork, pallet, and pallet pocket) in real-time. This can shave five seconds off every load moved assuming a camera frame rate of twenty hertz. This is valuable savings in time and money. There is also the possibility of deploying an advanced control system to affect robot-to-robot communication and cooperation leading to even more operational efficiency for this market application. The revenue model is not clear. Can it be per truck per pallet or maybe a per truck license fee? We don't yet know how to price it. I don't want to short-change my own company but how does a time saving of five seconds equate to dollars in a fair way keeping in mind our core tenet of delivering more value than you extract."

What do you see as the risks?

"I don't think we have too much technology risk, and although these are not easy problems to solve, it is well under control. The biggest risk is on the business side. There are half-a-dozen promising opportunities but if we don't close a large deal leading to recurring product revenues in the next six months we'd be in trouble. Contract work does bring in some money and we have angel investors to help extend the runway but ignoring the situation doesn't make it go away. We'd have to ask ourselves if we're solving the right problems, that perhaps the market isn't demanding this technology. I don't want to take money from NSF and angels for nothing without market validation of some kind in the next six months. There's a good chance we'll get two or three large contracts. On at least one of these the technology pieces we provide is critical to a product that company already has in the market. Another is the slow-moving one from Pride. If we close on one or two of these, the three angels may invest again. These will amount to more than a million dollars – sounds good but I must make it happen."

Financing

Starting funds were derived from the consulting contract Tom had with the company that purchased Freedom Sciences, and from the sale of his small equity stake in that company. Three angels invested $130K in exchange for equity of which Tom owns eighty-seven percent. He was fortunate that eighteen months into his new startup, in January 2012, he received a six-month Phase I award of $150K from the NSF SBIR program. This was followed by a two-year Phase II award of $500K that started in April 2013. "The NSF SBIR program with its timing and the funding amount almost serves as a roadmap. The program is mature, robust, and time-tested. There is value to how it's laid out. With its various supplemental incentives, it encourages startups to partner. For eighteen months, I could generate sufficient revenue through consulting work to think and work on a product. With the initial funding from NSF, I began to build a team around a product model and hired our first employee well-known to me. He had worked with me at Freedom Sciences and has equity in Love Park. It's not wise to take a risk with your first employee; now there are three fulltime employees and two part-time. It was during this Phase I that Pride contacted us."

What motivated angels to invest in Love Park?

"One of them, a businessman with manufacturing domain expertise, had previously invested in Freedom Sciences. Back then he told me – 'whatever you do, keep in touch, I'd like to keep tabs on you.' The second angel had recently sold his healthcare business to a big pharmaceutical company and has an interest in robotics. I have known him for a few years. One night he sent me a text message that said – 'Let's have breakfast; I want to invest in Love Park'. I never would have predicted that he would invest in my company. The third person is a professional and friend. He works at a local financial trading company, has a background in computer graphics and an interest in robotics. I used to run a local programming group until I resigned last year. We used to meet there once a month. One can never underestimate the value of networking; I am almost always networking."

Company Growth

Tom believes that in the robotics business every employee needs to be technical at this stage of the company. He maintains that there are professional service providers available to take care of routine business operations, and that he cannot afford to have a skill overlap in the current three-member team. Robotics is a broad area, and the mix consists of a computer scientist (Tom), an electrical engineer with a master's degree, and a mechanical engineer. The startup plans to be a ten-person team in two years, to add more depth, including a sales and marketing person, and a business development person.

Why choose to have a ten-person team?

"It's a function of the types, quality, and the amount of work driven by the existing product mix, and what other product initiatives the market throws up. It also has to do with my experience. I am comfortable with a ten-person team. Beyond ten people, the company changes quite a bit. With a small team, it is easier to keep them strong together, more focused, and it can be lots of fun. With a larger team, I feel it becomes more management oriented. To build the team I lean heavily on my network; I network at least a couple of nights per month, always looking for people, people flow is like deal flow. I prefer to hire a known person if possible."

"I maintain strong relationships with universities for smart graduate students. John at Lehigh can help. I fund John through SBIR funds. It will be fantastic if John can take a one-year sabbatical and work fulltime at Love Park. He has research goals that mesh with a few applications I have in mind. The current three-person team works well together. They have signed on to developing products that they genuinely believe will be useful to people. The company culture involves a loose structure, is cognizant of personal life needs. It is not a 9-5 job. If a team member sends an email at eleven in the night, for instance, there's usually a response within a few minutes from the other two. Working together is not a challenge. Our location is also a useful resource. We operate out of a 'makerspace', a 21,000 square feet facility with all required equipment like Computer Numerical Control (CNC) machines, water jets, 3D lasers, lathes, and a dozen other resident hardware companies. This also provides access to like-minded professionals and small contracts we can quickly fulfill to generate additional revenue."

In terms of the team, would you have done anything different?

"If I should do this over, I'd have a co-founder, someone to close deals, do sales and marketing, and business development. I am OK with identifying customers and partners but to land a customer, you must nurture the relationship, and understand their needs. It requires the same rigor as for technology development – I call it customer engineering. The business side of technology startups gets lower priority but is so critical. Also, with a co-founder you'll have two different perspectives."

Do you seek help from others?

"I am comfortable negotiating deals. I know enough to handle it myself. We are not selling anything yet. I am working on a licensing deal with Pride, an AGV-type contract with a second company and a few others so that we have recurring work. I lean heavily on the three angels – they act as my advisors. They have broad experience, interests, and insights - from startup to selling a company. When I reach-out to them, I feel I am not alone, that it's not the end of the world when something like the Pride deal didn't work out when we thought it would. They have been through all this; on dark days, just talking to them on the phone helps a great deal. I email and call them regularly and meet face-to-face once every quarter."

Key Relationships

"In our work with Pride, they laid out product-level milestones. Often some wonderful stuff that we may dream of doesn't matter. They tell us what they want, more product focused rather than fancy technology development. Even with their recent acquisition, our relationship is not dead. Pride remains a strategic partner."

Love Park has not yet licensed anything to Pride although that company provides product direction and market insights. Tom knows that prior to their recent acquisition, the relationship was tight and that it will soon come back. John and his graduate students at Lehigh are important. NSF and the three angels are critical on the financial side. The company's ability to work with 3D data, 3D cameras, object recognition, and a variety of sensors helped it develop multiple vendor relationships. Such vendors and suppliers have provided three recent important leads to product development opportunities and future contracts. The makerspace is also a key connection providing subsidized rent, 24/7 access to a huge, well-equipped facility, and the presence of other hardware companies.

"We bring people to meet us here with great confidence because our location makes us look like we are a much larger company. We meet with potential customers, partners, and investors in a beautiful boardroom, and they are impressed. It adds to our credibility. Pennsylvania is reinvigorating its manufacturing sector. This provides good connections to manufacturing personnel – there are training courses held, for example, in agile manufacturing and other skills."

The Startup Experience

What advice would you offer a young entrepreneur?

"You'd better love what you do; got to be passionate, and you must care. Beyond your family, that is what you are married to. You have to love your team - you spend lots of time with them. A startup is not a research project. It is a business, and business development requires equal rigor just as with technology development. When you work with a big company, you must be patient and must have time on hand for they are slow, and they don't move like you. While a startup is in the business of innovation, big companies are in the business of risk mitigation. When you are just starting out, you are unproven. Leverage your network to get "warm intros" to the right people. You should never have all your eggs in one basket. You may have a nice, big fish on the line, but you may never catch it. Also, I highly recommend that one has at least one co-founder just as invested in it as you are."

Can you describe a high- and low-point?

"They were both at the same time! In June this year, our work with Pride culminated in a demo in their boardroom with all key decision-makers present. Our product worked flawlessly in an unknown environment unlike in a lab in Lehigh or in our office. It was a real high, almost enough we thought for them to take out their pens and sign away checks. A couple of days later I had a conference call with the same group to discuss their decision to place our project with them on the backburner to focus on an acquisition. It was very tough for us, a dark day, when we assumed that we had nailed it. It was a sobering conversation with my wife that evening and with the guys the next day – very tough."

What is it that makes you want to do this?

"I am pragmatic. I love what I do - it is intellectually exciting and stimulating. I believe it will positively impact lives and help people. There could also be good money behind it that I can share with the team. Life is chaotic with two little kids – this work offers flexibility. It's a 24/7 job but at the same time I am free to go see my daughter's performance in a play at school or spend time with my sick kid. There's no boss and I don't have to ask anybody to be able to do this. I attach a lot of value to this. I also don't have to worry about the work – others will take care of it; I know they won't neglect their work, the work is going to get done; they will put in the time. In a larger organization, it's hard to have conditions and rules like this because people will game the system."

What about luck, family, and personal qualities?

"There is persistence, I know there will be failures and successes every day and that I have got to keep going. There are life-lessons I learnt from my dad, a blue-collar executive who made himself. He made it but never handed anything to me or my sisters. I remember when I turned sixteen and obtained a driver's license and wanted my own car. My dad insisted that I earn it saying he'd match every dollar I worked for. One must work hard. Smarts is fine but challenging work is essential. As for luck, time and place has something to do with it. If we had the Pride contract, we'd be shipping products now. Don't bank on luck, save it. For me personally, it's been a wonderful experience. It provides context and order to my daily chaos. The NSF SBIR program has been very positive for me, and I'd love to give back, even if a little."

Where was this startup at the end of federal government funding?

Equity investments and sales total approximately $1M. The technical focus has been centered upon two core products/technologies. The first was the Pallet Detection System (PDS), and the second was sensor interface software for a 3D camera product. PDS provides 6 degree-of-freedom pallet position and orientation estimates to automated fork trucks to facilitate a fast and safe pick. The PDS runs directly upon a 3D camera mounted on the carriage of a fork truck and is situated between its fork tines. The way the company expects to achieve scale is to not only market the PDS as a turnkey product to automated guided vehicles (AGV) vendors but to also pitch it to *manual* fork truck companies as acritical component to building Advanced Driver Assistance Systems (ADAS).

ADAS development in the fork truck space is a hot topic due in no small part to the focus and popularity of mainstream autonomous vehicle technologies. The PDS will be a critical enabling component for manual fork truck ADAS systems. One of the tools the company must help market and sell the PDS is an evaluation kit which provides a rapid means for potential customers to try out the PDS before expending any engineering resources on it. The PDS evaluation kit comes with all the hardware and software necessary for a customer to begin evaluating the performance of the PDS immediately. The final item is that the company is actively expanding the PDS to include not on the ability to "pick" pallets, but also to safely "drop" pallets.

Like many other hardware companies, Love Park is realizing that hardware, particularly in the rapidly moving 3D imaging market, commoditizes quickly due to the stiff competitive nature of the market. Further, the company realizes the value their hardware provides in the market is largely a function of the solutions that can be built upon their sensors. At the start of federal funding there was a single employee. The near-term goal at the end of this funding is to grow to six employees with a revenue target of approximately $1.5M. Most significantly during this federal funding, the company was able to successfully pivot its market focus from the electric-powered wheelchair market into industrial automation. Love Park has been able to leverage its expertise and the core technology developed in support of active wheelchairs to refocus it into a more business-sustaining environment – one with tremendous growth potential. Thus, the first important commercialization result is the "rebranding" of Love Park as a provider of next-generation industrial perception solutions.

Where's the startup now?

Love Park's technology stack is the result of decades of experience in developing 3D perception applications. It builds fast, intelligent applications that are designed for use in industrial environments. These high-performance applications minimize or eliminate dwell time during a typical cycle, maximizing productivity. This delivers several critical benefits, including high performance, clean integration within a robotics software stack, sensor interfaces, and application/algorithm development. Love Park solutions are characterized by a combination of open source, proprietary software, real-world industry experience, and systems integration.

Love Park's focus is in the development of next-generation industrial perception solutions. However, robotics applications are not limited to perception. Love Park has full-stack robotics experience, deep understanding of the interplay between subsystems, and works with customers on delivering a complete solution. Such applications produce a valuable "data exhaust" that can be streamed to clients' "industrial internet of things" platforms. Example areas where this data could be used include preventive maintenance and enhanced inventory management.

Case Study Questions

1. Strategic partnerships – aligning expectations; partner as competitor
2. Diversification – looking for opportunities in adjacent markets
3. Federal government role in human assistive technology development
4. Is the venture capital model applicable?

5. Develop an alternate business strategy and business model
6. The question of broader social impact and market-size
7. NSF SBIR Program route constraints if any
8. Comment on team build-out strategy and team size
9. Develop a growth and exit strategy
10. Identify risk elements and how to manage/mitigate them?
11. Will regulatory issues hamper or aid growth?
12. Take-away(s)

11   You Can Print Batteries?

Technology Sector: Energy Storage
Startup: Imprint Energy, Inc.
Website: www.imprintenergy.com
Location: Alameda, California

Federal Funding Timeframe: January 2012 – May 2017
Funding Amount: $1.31M

This narrative is a story of: The need for private equity funds beyond federal funding; contracts; licensing; joint development agreements; technology pivot(s); business model changes

Startup Formation

Imprint Energy ("Imprint") was incorporated in 2010. The three founders are Christine Ho, Brooks Kincaid, and Devin MacKenzie. Christine completed her Ph.D. in materials science and engineering from the University of California, Berkeley ('Berkeley') in 2010. Her father, an entrepreneur and CEO of a company he started when Christine was in Junior High, was her inspiration. She vividly remembers going to her first trade show with him and the pride he took in what he was doing. Christine frequently talks to him about his entrepreneurial efforts – in conversations with her father, she shares her own startup experiences, and finds it useful that he can directly relate to them. As a graduate student in Berkeley, she found great entrepreneurial spirit within the campus and within her own department where many peers routinely started their own companies, thus raising awareness, sharing stories. Christine is particularly interested in research outcomes that directly impact society and translating technology development to commercial products.

Devin, after completing his Ph.D. in electronic materials science, aimed to become a university researcher. In 1998, he stepped outside of his previous focus on conventional approaches to compound and silicon semiconductor devices and took a physics post-doctoral position at the Cavendish Laboratory, University of Cambridge, where he conducted research in the then new field of organic electronics. While at the Cavendish, he led work in organic photovoltaics and self-assembling organic light-emitting diode (OLED) materials. Devin also began research collaboration with Seiko-Epson which led to the realization of the fundamentally disruptive potential of these new electronic materials and print-based processing. Soon thereafter he co-founded in Cambridge, the world's first printed electronics company, Plastic Logic. Starting with that initial experience, Devin has now founded or restarted five startup companies in printed and flexible electronics. Devin first met Christine while at printed flexible OLED pioneer Add-vision. A couple of years after that introduction, they began to discuss directly collaborating in the printing of batteries. This interaction eventually led to Devin becoming CEO of Imprint.

Brooks has an MBA from Berkeley and an undergraduate degree in management science and engineering from Stanford University. When he met Christine, Brooks was an MBA student at Berkeley's Haas Business School. Brooks and Christine attended a 2010 spring course on research-to-market strategies for new inventions from Berkeley, Lawrence Livermore Lab, and other national labs called Cleantech To Market. It was a pleasant revelation that they had both graduated from the same high school. Through this course, Brooks and Christine determined that there existed sufficient commercial potential and market validation of Christine's Ph.D. work. This positive market feedback was sufficient reason for them to incorporate a company in December 2010. They realized it takes money to do hard science – Christine and Brooks entered business plan competitions. At first, they won awards of a few thousand dollars. They continued to win more such competitions, and in time these awards totaled approximately $50K, enough to get them started.

Technology

The core technology was researched during Christine's doctoral work at Berkeley. It began with the development of a robust, thin, high conductivity polymer electrolyte that enables the ability to recharge in zinc-based battery chemistry[17]. It uses non-lithium rechargeable electrolyte chemistry - this eliminates the safety and toxicity issues associated with lithium batteries. This electrolyte also enables low-cost, air-based printing of thin batteries on commercial printing equipment in flexible form factors. Current R&D efforts are focused on expanding the performance characteristics of Imprint's ZincPoly™ printed, flexible battery. This battery technology utilizes air-stable, earth-abundant, non-lithium materials that facilitates manufacturing using print-based processing and is scalable to large dimensions. The company is targeting the mismatch between available thin-format battery technologies and the performance, form factor, cost, and ease of manufacture for a platform battery system to power flexible and wearable electronics, internet of things, medical, and portable electronic devices.

Imprint and the university co-own the intellectual property (IP) generated by Christine's research at Berkeley and the company has exclusive control of the patent. The company has created substantial additional intellectual property based on Christine's core concept including new materials and processes.   No external IP is required to practice the technology. Process patents and knowhow are an important part of the technology, but the highest protectable value is in the material formulations and in the materials supply-chain. Both are valuable – for now the strategy is to protect the formulations in the form of patents and the making of inks and the associated material supplier groups as trade secrets. Key specifications, milestones, and timeframes on the technology development roadmap are determined through close interaction and discussions with potential customers and partners on their needs and requirements. Battery specifications are application driven.

Specific technical solutions to more advanced products, and their related manufacturing processes, are currently not envisioned. Technical advice if required is currently ad hoc - the company approaches Christine's Ph.D. advisor on such matters. The use of external consultants is sparse – it is institutional, and mostly from academic research centers. The company plans to soon constitute a formal Board of Advisors. There is rarely a need to approach experts with a single question - it is more about developing and sustaining long-term relationships. In 2010-11 when it was getting started Imprint used the nanofabrication lab at Berkeley for technology development. This lab allows many startups to access its facilities for a monthly fee, and the startups get to own any IP that is thus generated. Imprint now has its own printing capabilities and wet- lab facilities, research tools, instruments, and equipment, to develop all basic functionalities and materials formulations in-house.

Customer Discovery and Business Model

The management team, customers, investors, and the Board work together to help predict this fast-evolving technology market. This motivates product development especially since many possible products do not exist today. In some ways, the startup has to create markets and develop a strategy for a select few customer segments. The Board guides the company along the right pathways. When asked about finding potential customers, the founders had this to say – "It's not hard to generate interest in what we do - lots of customers have come directly to us looking for a solution; too many enquiries, more than we can handle. We have potential customers from across the board – the problem is to choose wisely."

What kind of customers?

"O, wearable space, athletic and fitness monitoring, consumer electronics…big advantage to this technology, as it is non-hazardous with low toxicity, customers in transportation, disposables, medical devices. The press writes about us – this generates lots of interest. We go to conferences regularly; have an exhibit booth or make a presentation – lots of traffic; some want products to put out into the market now, others are interested in wearable electronics, others in developing new products they can potentially design with our technology; any of these are interesting initial starting points but not the endpoint. We'd like to have multiple partners, multiple products – often educating them on what we can do."

Imprint now plans to focus on a small set of lead customers/early adopters to seek a design win in the next eighteen months followed by pilot manufacturing of their first products. They will down-select, using a set of pre-defined, but adaptable, set of criteria such as market opportunity and new product specifications; if partnering will lead to more foundational work for a bigger opportunity later; will it advance the technology, process, and customer supply-chain; will partner be willing to collaborate with a small company over the long-term. The company has a close working relationship with initial customers and two select partners in the U.S., Asia, and Europe in different parts of the supply-chain such as material suppliers, manufacturers, and end-designers. It needs to be selective and simultaneously be open to varying market dynamics. It would be too risky to bet on just one.

Although revenues are being generated through joint development agreements (JDA) and non-recurring engineering (NRE) contracts, the real value is in the flexible battery materials leading to a low-cost manufacturing process. The company has formal agreements with partners through JDA and NRE projects with timelines and deliverables. These may morph into licensing agreements. They also have an investor agreement with a large manufacturer. Eventually the business model will evolve into a combination of being a supplier of materials/ink and a battery technology licensor whereby the company helps customers design their product which will in turn be mass-produced by a contract manufacturer. Materials are formulated in-house, and customers are sold containers of different inks and paints that are then converted to flexible batteries of certain form factors and performance specifications. This outsourced manufacturing model will involve multiple partners, multiple customers, and multiple manufacturers.

The biggest competitors are Asian manufacturers of battery cells for small portable devices using conventional lithium polymer battery chemistry. This is the incumbent technology currently used in most portable, rechargeable electronic devices. For thinner, flexible electronics, this technology has known limitations related to toxicity and safety. Manufacturers are aware of these but are used to producing these products for over a decade. Potential small portable electronics customers evaluating these technologies have determined that the limitation is one of power density not the usually considered energy density. In wireless, mobile applications, power requirements are constraints, and the defining problem. Many wireless technologies such as Wi-Fi and Bluetooth desire batteries with low internal resistance that can provide adequate power to wirelessly communicate. Niches exist in the small portable electronics and consumer space, each with their unique set of performance specifications – the wearables space is attractive but has rigorous and the hardest-to-meet specifications whereas performance specifications are easier to meet in the wireless sensor and the Internet of Things (IoT) space.

This is a new battery technology with new chemistries and new form factors, and so it is critical to build trust with customers. This takes time and time is something a small startup cannot often afford. Many market drivers co-exist – with wearables there is growing momentum and the need for IoT apps are also increasing –

"We need to recognize customer market pull – shift our focus as required; we need to be nimble, and we need flexible early customers."

Financing

Imprint was officially incorporated in December 2010. From the initial stages, all three founders, fully committed to the long-haul, provided sweat equity before they began to win business plan competitions and grant awards. The first award from a Berkeley campus competition garnered $2K with which they could purchase a screen printer. Then awards from higher profile national and international competitions brought in bigger checks that totaled $40-50K. In 2011, the company was awarded two grants, one of which was from the National Science Foundation (NSF) Small Business Innovation Research (SBIR) program. It also obtained grants from an internal grant from Berkeley. Currently the NSF SBIR program remains the only public source of funding. At first it was challenging dealing with NSF – there was a delay of six months in the disbursement of Phase I funds ($150K), then there was the government shutdown, but this funding has always been essential to the founders' ability to grow the company. In general NSF support was timely, useful, and matched company growth - "The NSF SBIR supplements are incredible – very important and helpful; one helped us get a business partner to work with us just at the right time, another was great for hiring interns in the summer. There's another one that would allow us to hire a post-doc – we could use that."

Convertible debt from Dow Chemical and later support from In-Q-Tel in the form of a research contract and equity warrants provided runway of two years. Revenues came from three sources – a third from grants, a third from customer NRE contracts, and a third from Dow-like entities. The team began to realize that it was not prudent to continue to scrape by as a seed-funded entity - they needed to grow, quickly solve technical problems, accelerate the technology development, and secure IP. The wearables market and the IoT market were primed for take-off - the company needed to be better-positioned, with additional resources. Raising capital to finance a battery technology company is difficult and takes time – investors, more comfortable with software startups, are wary of battery companies. The well was poisoned too - battery companies had gone bankrupt, there was investor-prestige risk, to them hardware and materials startups posed too much risk. Against all odds, the company closed on a $6M investment in May 2014. Phoenix Ventures was the lead investor with two follow-on investors – Flextronics and AME Cloud Ventures. The investors found the Imprint team credible – they liked the team which was most important as they did not suggest changes to the management team. The team believes this $6M investment is enough to get them past a commercial design win.

How was the size and valuation for this round decided?

"We started financial planning last year – the number of people to hire, their profile, salary, overhead – these were the main drivers; commercial timelines and applications were two other drivers. We discussed matters with future and current investors – whether the strategy, budget and timelines were reasonable. Regarding valuation, we had an optimized number in mind; the investors had theirs – there was some back-and-forth based mostly on risk assessment, market timing issues, deliverables, and milestones; both sides compromised somewhat."

Do the new investors have domain expertise?

"Phoenix Ventures only invests in advanced materials companies. The general partners were entrepreneurs, all have operational experience – they have networks and an understanding of what it takes to build a battery company. Their limited partners are large chemical and manufacturing companies, and they in turn have strategic partner relationships with Asian manufacturers, and partners in the U.S. and Europe. We feel they bring credibility and value."

This seems to be smart-money; would it be patient-money too?

"They have a practical understanding of this market segment. They too are under pressure to succeed – they maintain an active materials portfolio; some of their previous investments were successful exits."

Company Growth

As of now, there are eight full- time employees soon to grow to twelve. All employees have equity shares in the company. Additionally, there are student interns, some of whom are supported by NSF SBIR through various supplemental funding mechanisms. The $6M investment is spread over two years with additional funds expected from grants, and NRE/JDA work. These revenues currently total about $1M per year. The founders periodically meet to discuss new operational structures and are in the process of codifying them.

Over the long-haul, Christine is interested in the learning process, in the journey, and maintaining positive relationships both internal and external to the company. The founders desire to build a company that people are proud to work for.

Devin believes in success, to make a social and technology impact, and is not interested in merely publishing papers like most other researchers. To impact people and society, he says, the company must be commercially successful. Brooks, the business development person, would like to be open-minded about the technology and its outcomes. Decisions are made by the management team with guidance, direction and help with strategy from the Board consisting of five seats – Devin, Christine, and two lead investors; the last one remains vacant. The Board helps with introductions to potential partners, and Board meetings every other month with monthly updates usually involve extensive potential partner discussions and next steps with current partners. Asked about founding team dynamics and equity distribution, Christine said –

"Brooks and I have built a good working relationship over the last four years – we were the two original founders with Devin joining us a year later. Brooks is business development and I'm technical. Initially, we spent considerable time learning each other's language – educating each other; how to tell our story, learning each other's vocabulary. We have together become more and more credible to investors, partners, and customers. Devin has both a technical and business background with five different previous startups. He spends half his time with me, the other half with Brooks; having Devin on the team helps greatly with credibility and experience; for Brooks and me, this is our first startup. Equity division is relatively even – Brooks and I started a year earlier, so we have a little more equity; together the three of us own a large majority."

A few key company personnel manage and execute large pieces of the technology development effort. These include a senior person with a Ph.D. and experience in the battery industry, two others with master's degrees, and an expert printer with thirty years of experience in the printing industry.

What if a key person decides to leave?

"It will be tough. We incentivize them with equity and a bonus program that includes a review system with clear expectations to reward work above and beyond the norm. It also provides better insights into how they are doing. Everyone does a lot of things – even when someone goes on vacation it is painful. We try to cross-train and have recently instituted a 'buddy' program – if one designs an experiment, we ask that he runs it by another; we try to be open and transparent; we try and keep everyone excited about what we do."

How about mistakes made hiring people?

"It is difficult to find the right people - much time spent time on recruiting, finding people with the right skills, people with experience in industry. We try to leverage our networks, use LinkedIn, takes a lot of work, over 300 resumes, 40 phone interviews, and 5 in-person interviews just for 1-2 additional hires. When we first started, we looked for a specialist with specific domain expertise, didn't then account for strategy changes or delays; not all specialists are adaptable to fast-changing situations. We'd like to stay small and lean; we don't expect exploding growth in the next 3-5 years. We plan to use partners to offload work, and we'll have no in-house manufacturing."

Key Relationships

The founders believe that the NSF SBIR program and Berkeley came through with help and funds at critical times. They are fully aware that materials and hardware companies take time and patient money to succeed. Supplemental funding from NSF SBIR helped with hiring interns, with securing strategic partnerships, and with trade show participation in Eureka Park at the Consumer Electronics Show. Imprint regularly participates in major consumer electronics trade shows – these and consumer electronic designers have led to many potential customers. Technical conferences are mostly to attract future employees and partners. Spreading the word with advertisements, positive press, and speaking at Berkeley campus events also helped. Licensing negotiations were critical and difficult at times – it's not easy for anybody, but both sides were committed to success.

Strategic partners, lead customers, and early adopters have been important for technology development, helping to craft a roadmap, and with revenue generation through JDAs and NRE contracts. Investors have made timely introductions, and partners such as Dow and Flextronics have assisted with manufacturing insights, and potential customer channels. Material suppliers like Solvay, one of the biggest chemical companies in Europe, are useful to develop new materials and IP together. Imprint has a JDA with Solvay, is utilizing their materials database for materials property optimization, and might even scale up ink production with that company.

The Startup Experience

Christine was asked about her startup experience

"It's been an amazing journey, I've enjoyed every step, the lows, and highs; it feels incredible. It's exciting, lots of lessons learned along the way, and it is a great privilege to develop a unique energy-related platform.

People see our videos and press articles and are inspired by what we do; this makes it worthwhile; it has been a fantastic opportunity to learn and pass knowledge on, help the next generation."

How would you compare working in a large multinational corporation, a famous university, and a startup?

"I don't think about it that way – I'd have been equally privileged to extract value and knowledge, all three are great, I'm confident I would have learned so much in each case. This opportunity was presented to me, it was a matter of following through – risk, career, money, I did think of it, and it was a difficult decision; starting a company seemed most attractive at that time, and I was willing to take the risks, I have no regrets, this is what I am doing now, and don't see the other options as lost opportunities."

"I have one piece of advice – in grad school, I worked in the lab a lot and acquired many skills, but it is most important to build relationships, it can often be lonely. If you build relationships, your network can affect your future, your career. Many people approach me, want to know my story, hear about my experiences – some have practical questions – about startups, licensing; I want to pay it forward by helping as much as I can. This is really my first job; it's funny and strange recruiting people when I myself have never been recruited!"

How do you judge people?

"No easy answer – it's about intuition and gut-feel as well as research; I even experience that in a social setting – seeking feelings of trust, energy, synergy; the business world is even more complex –it helps to collect as much data and check references."

On personal attributes and what motivates you?

"A combination of many things – from the stress management perspective, I can handle stress, can separate work stress from personal; this has evolved over time; keeping cool is important; I am an optimistic person, and tend to see things from a happy point of view – this is not being naïve, but I often give people the benefit of the doubt. I plan for success – a commitment to move forward, drive to keep momentum going, don't like to stop or sit still."

"I'm interested in solving problems – if things don't work, there must be another way – persistence and positivity; fame and money are not so motivating, it is success and accomplishment. The biggest learning is the power of working together, working as a team to do amazing things, 2+2 is not 4! What drives me is harnessing and building relationships and teams to achieve remarkable things together; groups are powerful. When you align, and motivate a team, your team can do amazing things."

What about sheer luck and good fortune?

"A lot of it feels like luck – meeting Brooks, ten years after high school, it immediately bonded us. And meeting Devin – it was seven years ago, at random, through an advisor I worked with in Japan for a summer who was a post-doc friend of Devin's at Cambridge. Years later I contacted Devin asking for startup advice, and he is now the CEO of our company! Many meetings with investors and customers came from chance meetings. Recruiting our first employee, how we met and connected was by chance, and he became such a big part of our company and whatever success we've had so far. All of this seems lucky."

Where was this startup at the end of federal government funding?

Imprint was able to unlock the second tranche of their Series A investment ($2.56M) by reaching commercialization milestones, including securing a design win with a customer. They also signed a license agreement with a manufacturing partner, sold battery samples for revenue, and secured multiple joint development agreements and non-recurring engineering projects with partners and suppliers for a total of approximately $4.5M. There are no products in the marketplace, and no patents resulting from the SBIR project yet. At the start of federal funding there were two employees; the company now has twelve employees. The NSF SBIR award was pivotal to enabling Imprint to hone its technology, refine its business model, and provide a pathway to engaging with customers. The data and capabilities generated through this project have been critical to capture customers and partners. Customer feedback drove some of the milestones and business strategy pivots captured throughout the project tasks. Imprint's Series A investors viewed the series of awards from the NSF throughout the SBIR project as validating evidence of Imprint's capabilities and commercial relevance.

Where's the startup now?

Wearable devices and mobile sensors would be improved with power sources less bulky and rigid than existing batteries. This California startup is developing flexible, rechargeable batteries that can be printed cheaply on commonly used industrial screen printers. The company has been testing its ultrathin zinc-polymer batteries in wrist-worn devices and hopes to sell them to manufacturers of wearable electronics, medical devices, smart labels, and environmental sensors.

The company's approach is meant to make the batteries safe for on-body applications, while their small size and flexibility will allow for product designs that would have been impossible with bulkier lithium-based batteries. Even in small formats, the batteries can deliver enough current for low-power wireless communications sensors, distinguishing them from other types of thin batteries. The company recently secured $6M in funding from Phoenix Venture Partners, as well as AME Cloud Ventures, the venture fund of Yahoo cofounder Jerry Yang, to further develop its proprietary chemistry and finance the batteries' commercial launch. Previous investors have included CIA-backed venture firm In-Q-Tel and Dow Chemical. The batteries are based on research that company cofounder Christine Ho began as a graduate student at the University of California, Berkeley, where she collaborated with a researcher in Japan to produce microscopic zinc batteries using a 3-D printer.

The batteries that power most laptops and smartphones contain lithium, which is highly reactive and must be protected in ways that add size and bulk. While zinc is more stable, the water-based electrolytes in conventional zinc batteries cause zinc to form dendrites, branch-like structures that can grow from one electrode to the other, shorting the battery. Ho developed a solid polymer electrolyte that avoids this problem, and provides greater stability, and greater capacity for recharging.

Brooks Kincaid, the company's cofounder, and president, says the batteries combine the best features of thin-film lithium batteries and printed batteries. Such thin-film batteries tend to be rechargeable, but they contain a reactive element, have limited capacity, and are expensive to manufacture. Printed batteries are non-rechargeable, but they are cheap to make, typically use zinc, and offer higher capacity. Working with zinc has afforded the company manufacturing advantages. Because of zinc's environmental stability, the company did not need the protective equipment required to make oxygen-sensitive lithium batteries.

"When we talk about the things that constrain us in terms of the development of new products, there's really two that I lose the most sleep over these days. One is batteries and the other is displays," says Steven Holmes, vice president of the New Devices Group and general manager of the Smart Device Innovation team at Intel.

Despite demand for flexible batteries, Christine says no standard has been developed for measuring their flexibility, frustrating customers who want to compare chemistries. So, the company built its own test rig and began benchmarking its batteries against commercial batteries that claimed to be flexible. Existing batteries failed catastrophically after fewer than a thousand bending cycles, she adds, while Imprint's batteries remained stable. Imprint has also been in talks about the use of its batteries in clothes and also, according to Christine, "for weird parts of your body like the eye." The company also recently began working on a project funded by the U.S. military to make batteries for sensors that would monitor the health status of soldiers. Other potential applications include powering smart labels with sensors for tracking food and packages.

Case Study Questions

1. Comment on hardware startup companies in general, and specifically on battery technology startups
2. Assess commercial potential and value proposition
3. Appropriate business strategy/business model to adopt?
4. How best to fund this endeavor? Why NSF SBIR Program? JDAs and NREs to generate revenues; VC money - smart-money, patient-money?
5. Start-up valuation exercise
6. Founding team dynamics – ideal number of founders
7. Identify risk elements and how to manage/mitigate them?
8. Go-to-market strategy and properly timing market entry
9. Growth and exit strategy
10. Partner relations – communications/clarity; expectations alignment
11. Take-away(s)

12 Down with Internet Hackers

Technology Sector: Cybersecurity
Startup: BitSight Technologies, Inc.
Website: www.bitsighttech.com
Location: Cambridge, Massachusetts

Federal Funding Timeframe: July 2010 – June 2014
Funding Amount: $1.15M

This case study is a story of: A garage-to-market innovation model: federal lab served as technology source; private equity funds; rapid growth; acquisition of smaller company for synergy/ growth

The Garage-to-Market Innovation Model

Although technically the genesis of the technology developed by BitSight was in Lincoln Labs, the point to be made is that most software companies can have their start in a garage as is the case with this garage-to-market innovation model. This model can further be highlighted by examples such as Google, Amazon, Facebook, and Microsoft all of which essentially started in a garage. This is especially true in this day of the so-called "app economy", where an idea, time, a bit of funding, a few computers, and code-writing skills is all that is required to get one started. It doesn't require a large infrastructure, a suite of test, validation, prototyping and pilot manufacturing capabilities, and scale-up of production as would be required in the manufacture of hardware, things tangible. This is differentiated by the lab-to-manufacture innovation model highlighted by another case study that requires the deployment of our country's manufacturing heritage to rebuild global competitiveness, thus, to allow startups to develop their businesses from the lab-scale prototype stages of innovation to the later stages of production and commercialization.

Areas such as functional thin-films, energy storage, and biomaterials - all critical elements of the nation's future advanced manufacturing economy – would require resources beyond computers and floor space. An experienced workforce with skills to develop and manufacture future technology products is essential. To support such new technologies, risks must be removed in order that the predictability of product maturity from concept to full-scale production be increased. We would need to provide the infrastructure for manufacturing innovation to ensure that the next generation of processes and products not only will be invented in the U.S. but scaled up and manufactured in the U.S. as well. This is both critical to the nation's economic future as well as to solving some of the world's biggest sustainable development challenges at this crucial juncture.

Company and Team

BitSight Technologies ("BitSight") was founded by Stephen Boyer and Nagarjuna Venna after their grant submission to the National Science Foundation (NSF) Small Business Innovation Research (SBIR) program was awarded in 2010. The origins of the company can be traced back to 2008, when the two founders were part of a team that competed in the MIT 100K Entrepreneurship Competition as an enterprise security software entity that aimed to commercialize the NetSPA technology developed at MIT Lincoln Labs. Thereafter, in meetings with potential customers and partners, the founders determined that a better market opportunity exists if they can successfully address the third-party vendor management problem. This led to the idea of building an information security rating marketplace that could introduce to the vendor management space the same efficiencies that Fair Isaac Corporation (FICO) scores have established in consumer financial markets.

Stephen was a Technical Staff member in the Lincoln Labs' Cyber Systems and Technology Group where he led R&D efforts involved in solving difficult national cybersecurity problems. Prior to joining Lincoln Labs, Stephen worked at a startup venture, Caldera, where he designed, developed, and tested a variety of products ranging from modifications and enhancements to the Java Virtual Machine (JVM) to the development of diagnostic tools for a distributed software and configuration management system. Stephen has an undergraduate degree in Computer Science from Brigham Young University and a master's in engineering and management from MIT.

Prior to BitSight, Nagarjuna worked at EXFO as a product manager successfully marketing and selling service assurance solutions to tier-1 carriers and service providers worldwide. He was responsible for managing the product lifecycle, defining the overall strategy and roadmap, and working with third parties assessing partnership and licensing opportunities. In addition, Nagarjuna was involved in developing corporate strategy to enable this traditional hardware company to increase its share of software revenues by becoming a strategic vendor to its customers. He holds an undergraduate degree in Computer Science from the National Institute of Technology, India, and a MS in Engineering and Management from MIT.

In 2012, Shaun McConnon joined BitSight as the CEO. The company's Board of Directors includes representatives from investment firms Commonwealth Capital Flybridge Capital, Globespan Capital, Menlo Ventures, and Michael Duffy from OpenPages.

Market Opportunity

No credit scores analog for information security assessment exists. Currently there is a highly fractured, non-standard, and inefficient assessment process that effectively requires each organization to be its own information security watchdog. If it wants to enter a business relationship or reevaluate the information security practices of one of its service providers, the organization must do so itself or hire an outside firm to perform the assessment for them. Carrying out separate assessments for each such relationship becomes time and resource intensive especially for entities and service providers with tens, hundreds, or even thousands of relationships. Additionally, most organizations are reluctant to share information on security related incidents. The resultant information asymmetry makes them reluctant to enter otherwise beneficial business relationships.

Specifically, when businesses share sensitive data with their partners, they face the prospect of being breached through these partners. Sensitive data may include customer information, intellectual property, social security and credit card numbers, usernames, and passwords. Today, businesses have limited means to manage this third-party risk proactively, efficiently, and effectively. Third-party vendor risk is currently assessed using information technology (IT) and security audits or assessments. These mostly annual audits are time consuming and expensive and are based on established policies rather than actual outcomes. The data from the audit, however accurate, becomes quickly outdated. The business finds it difficult to know, for example, if a partner is impacted by vulnerabilities announced after the audit is finished. The security policy may suggest that the partner will deal with the vulnerability but there is no straightforward way to validate it.

Due to the time and cost associated with audits, businesses only audit a small percentage of their partners. For instance, financial institutions audit at most 25% of their partners each year. In other industries, this percentage is even smaller. Thus, in many cases, companies are effectively blind to the risks posed by their partners. Not knowing if they are at risk because a partner's security posture is deteriorating is a substantial pain-point for risk management professionals. The time-intensive, costly, manual, once-a-year audits do not enable proactive risk management. As more and more services are outsourced or accessed through the cloud, the risk that a business can be breached through a partner has significantly increased. Businesses require a solution that can provide continuous visibility into their partners' changing security risk postures, so they can proactively manage the assumed risks.

BitSight's Continuous Risk Rating Service (CRRS) enables business to manage vendor risk efficiently and effectively. Ratings are generated daily based on the quality of actual measured security outcomes that result from information security practices at a given organization and are designed to be credible, predictive, scalable, and automatable. Just as the credit rating of individuals is continuously computed without any effort on their part, BitSight computes information risk ratings for an enterprise and all its partners. The service provides businesses with round-the-clock visibility into their partner networks enabling them to quickly respond to their changing risk profiles and enables comparisons of ratings across organizations. The service is accessible through a customer web portal where risk management professionals can monitor, assess, and mitigate partner risk using up to date BitSight ratings.

Using CRSS, enterprises can conduct risk-based audits rather than time-based audits. Using BitSight's service, businesses will reap the same time-and-cost savings that lenders derive from credit scoring services. These quantitative risk measurements enable risk pricing and focused mitigation strategies. Enterprises will also be able to perform apples-to-apples comparisons across partners. Complemented by on-site security assessments, the rating service can drive the enterprise specific decision-making processes in relation to third-party risk.

Technology and Product

The BitSight security rating platform generates objective, outside-in ratings on the security performance of companies. Using evidence of security outcomes from networks around the world, the company applies sophisticated algorithms to produce daily security ratings. The platform gathers terabytes of data on daily security outcomes from hundreds of sensors deployed across the globe[18]. This data is externally available and collected without any intrusive testing. Data is classified into several risk categories, including botnets, spam, malware, unsolicited communication, distributed denial of service (DDoS), and system configuration. These are then mapped to an organization's known networks. Algorithms analyze the data for severity, frequency, duration, and confidence, to create an overall rating of that organization's security performance. Security ratings, ranging from 250 to 900, are like consumer credit scores, with higher ratings indicating better security postures. New ratings are generated daily and presented via the company's security rating customer portal, for both individual companies and customer-created groups. Industry indices and historical ratings enable benchmarking and trend analysis. Alerts are generated upon significant changes in ratings, and actionable information is provided to mitigate risk. This platform can be leveraged for multiple applications such as third-party risk management, cyber insurance, and benchmarking.

As the company adds new data sources that provide new security-related events, the number of events that must be processed will increase. This massive increase in volume presents an ever-growing challenge to scale processing and delivery. Each element in the backend-processing pipeline must scale to ensure the entire pipeline can continue to run in a small fraction of the 24-hour cycle time currently in use. With the increased volume, the costs associated with the computation and storage has also increased dramatically. Costs must remain reasonable for a sustainable business. BitSight is developing new distributed approaches to data storage and algorithmic computation to assure that additional data and analytical computation remain cost-effective.

The generation of daily quantitative BitSight ratings for many entities is a complex process. There are innumerable parameters and variables that must be addressed and analyzed to provide ratings that have intuitive, actionable meaning to users who may not be technically sophisticated. The ability for its customers to have access to the ratings data to which they subscribe is an element of the minimum viable service that the company can provide. Not only must the customer be able to see their own rating and details of the security events that are attributed to their company, but they must also be able to review the ratings for their named business partners.

As the primary way for customers to interact with BitSight, the customer portal is enhanced on an ongoing, frequent basis. These enhancements include more quantitative ratings, the ability to present new data of interest to customers, and usability and form improvements. Customers appear to highly value the ability to compare a company's rating to that of others in the same industry in a time-series graphical representation. They also value the ability to display company and industry data by risk category. Another customer requirement is the ability of the service to alert them when the rating of one or more companies they subscribe to changes in a meaningful way. The notion of a portfolio risk is one that customers understand and value.

Several new categories of data such as fraud, DDoS, and undesirable employee activity that are useful in rating companies can likely be available to BitSight. Incorporating each of these will require more than simply collecting such data. It must be modeled to determine how to properly add it as a factor in the rating process and which metadata needs to be passed through the processing pipeline. Given the nature of some of this data, automated curation is also required. Today, all BitSight's event data is collected by and received from third parties. This has allowed the firm to focus on extracting the greatest business value from this data without the distraction of operating a data collection infrastructure. However, preliminary experimentation and analysis seems to show that certain types of data are not available through the existing third-party data vendor network. The company is therefore considering cooperative arrangements with various organizations where a BitSight network sensor system may be deployed to collect specific types of data that would be of value in the data analytics and ratings processing subsystems.

Business Model and Execution

The company's business model targets the existing audit-driven vendor management system with a solution sold using the Software-as-a-Service (SaaS) model - customers buy annual subscriptions to monitor their partners. It focuses almost exclusively on large companies in highly regulated sectors, including financial services, healthcare, and e-commerce. BitSight's technology and service has proved to be of excellent value in the financial services sector where it has two large customers and several other potential prospects. In general, the trend in the financial services industry is towards risk-driven vendor assessments rather than time-based audits. The total number of partner relationships in this industry is approximately three million of which about 15% are considered critical relationships and therefore prime targets for the startup's service. The firm signed a contract with a global retailer to provide vendor risk ratings on their e-commerce partners and is in discussions with an electronics payment processor. Healthcare companies have also shown interest in adopting this security rating product.

A key marketing and sales objective is to understand how companies in the e-commerce and healthcare sectors will use BitSight's service, what their main pain-points are, how much are they willing to pay, and if any changes are required in the service to meet the needs of these sectors. BitSight's biggest barrier to entry is brand identification. Currently, many financial service firms either use well-established auditors or conduct audits themselves. To convince them to stop using established brands they must believe that BitSight ratings are superior to those obtained from regular audits. While security professionals are aware of the limitations of the audit process, it is challenging to convince decision-makers to replace a major audit firm with a rating from BitSight. The firm's strategy is to identify early adopters in the target industries and work with them to develop successful use-cases. Early adopter customers are generally those who are concerned about being breached through a partner, are willing to work with new-technology companies, and are willing to try innovative approaches to risk management.

In terms of financing, the total NSF SBIR equity-free funding to BitSight in the period 2010-2014 has been $1.15M. The company closed a seed equity financing round in 2011 when investors bought $1.1M in preferred stock. A total of five investors participated in that round with the two principal investors Flybridge Capital Partners ('Flybridge') and Commonwealth Capital Ventures ('Commonwealth') accounting for $1M. In 2012 after Shaun McConnon came aboard as the CEO, and to accelerate development and hiring, additional investments by way of convertible promissory notes were made in the amount of $2M. These notes converted into preferred shares without discount at the time of the Series A financing round. Full-time employees of BitSight now total thirty-two.

In 2014, the company purchased Anubis Networks, a Portugal-based security intelligence company. This acquisition helps deliver unique data quality, breadth, and innovation to BitSight's security ratings. The acquired company's technology was incorporated into the BitSight platform and serves as a key data provider that drives the company's world-class ratings. The integration of Anubis Networks extends BitSight's position as the leading provider of information security ratings for organizations around the world.

BitSight's exit strategy is through an IPO or acquisition by a large ratings or risk management firm such as Dun and Bradstreet, Moody's, S&P, or Thomson Reuters. A successful IPO exit is possible if the company can establish itself as the de facto platform for information security ratings. If the firm has some success but is not the de facto standard, it becomes an attractive acquisition target for large ratings firms who may be looking to expand their portfolio of ratings to include information security.

NSF Commercialization Assistance and Impact

Innovation Accelerator (IA) founder and BitSight CTO, Stephen Boyer, were introduced at the MIT $100K Entrepreneurship Competition, where IA's founder serves as a judge. It was through this competition that the two men developed a professional relationship. In 2009, IA contacted Stephen to offer him the opportunity to be the lead developer in a project to address a specific security-related pain-point identified by an IA corporate partner. After several months of independent research and discussion between Stephen, IA and the IA corporate partner, the startup team decided, at IA's urging, to apply for a NSF SBIR grant.

Upon receipt of the grant, IA conducted a needs analysis and found that its assistance was required in the following areas - company structure; team building; product focus; market feedback; pilot testers; investors; and recruiting Board members and technology advisors. During this initial stage, IA was also instrumental in advising the company to reject a potentially toxic angel investment.

IA then focused on assisting the startup in their commercialization process: connecting to domain experts in cyber-security, compliance, and defense; facilitating introductions to fellow SBIR grantees for potential collaboration opportunities and guidance; managing relationships with corporations in the IA network who have expressed interest in BitSight; and introducing the startup's solution to the venture capital, investment banking, and security compliance communities. After the company obtained a NSF SBIR Phase II grant, IA provided additional assistance to develop a more focused business plan and market entry strategy; determine the optimal corporate structure and help form a powerful Board; critically analyze the failure of a competing product; help determine IP strategy; integrate into existing efforts in the cybersecurity landscape; and identify strategic partners.

One specific point of interest is focused on ways in which one such strategic partner, Continuum, would integrate technology developed by BitSight into their core business solutions as the startup product's market readiness matures. Continuum expressed support and worked with the company to further develop the technology, the scoring methodologies, and to test and validate its solution. Through IA, the company also built a relationship with the Center for Internet Security (CIS), a non-profit enterprise that helps large enterprises in the risk reduction of business and e-commerce threats. BitSight has interfaced with several large investment banks through their affiliation with CIS and continues to leverage the CIS network.

IA's assistance in the early execution and growth strategy revolved around BitSight's ability to develop their scoring methodology with as many strategic partners as possible, to be able to identify key personal and executive guidance, and to build a viable company with sufficient revenues to attract a strategic investment and/or partnership. IA continued to assist the company in its efforts to harvest influential contacts and customers, to guide in operational build-out, and to provide strategic support to the point of commercial self-sustainment. IA and BitSight have built a strong foundation together, their roots intertwined in an ever-evolving relationship, an outgrowth of a simple concept that has turned into an innovation with commercial potential and supported by the NSF SBIR program.

In January 2011, the inaugural "IA @" event series was held on in conjunction with the "Nuts & Bolts" entrepreneurial course at MIT. Three NSF SBIR-funded startups were selected to present as innovation 'case studies', BitSight being one of them.

Testimonial

"The Innovation Accelerator team introduced us to the National Science Foundation SBIR grant program and subsequently to a venture capital firm who financed our first round. The introductions proved tremendously valuable, but the IA team offered more than introductions alone. The collaboration helped us to identify the opportunity and formulate and refine our ideas. The quality of the work and the commitment of the IA team members during our interactions have been outstanding."

IA activities and outcomes are summarized in the following Table.

IA Activity	Outcomes
Startup Overall Engagement	18-24 months; 200+ hours of interaction; 250+ e-mails sent to and on behalf of BitSight
Recruit Management Team	Addition of co-founder Nagarjuna Venna
Form Partnerships	With fellow NSF SBIR-funded startups in the cybersecurity space – JAAL; Secure Banking Solutions; Stop-the-Hacker; and Thousand Eyes
Negotiate Commercial Deals	Goldman Sachs; Mutual of Omaha
Introduce Potential Customers/Partners	PayPal; GISC; Gallup; FBI; Northrop Grumman; Mastercard; First Data; TD Ameritrade; IPG; Fishnet; Solutionary; Booz Allen Hamilton; DHS; Stratcom; Microsoft; Symantec; Metaflows; Secure Banking Solutions; Ksplice
Find Investors	Introductions to Fairhaven Capital; Rustic Canyon Ventures; InQTel; Chart Ventures; Commonwealth Capital Partners; Allen & Co.; Flybridge Venture Capital; North Bridge Venture Partners; Southern Capitol; New York Angels; TechCoast Angels; Joel Mesznik. in 2011, raised capital including NSF SBIR matching funds totaling $3M from Commonwealth and Flybridge
Identify Domain Experts	Jeff Schreiner, former CEO of Continuum Worldwide; Kostas Mallios, Microsoft; Eric Russo, Symantec
Provide Proposal Guidance	Through the Phase I award process; assistance in Phase IB and Phase II proposal preparation
Feature at Trade Shows and Special Events	IA-sponsored Cybersecurity workshop in Omaha, Nebraska, to connect a select-group of NSF SBIR startups in the cybersecurity space with local corporations, federal agencies, and individuals focused on cybersecurity solutions, with opening remarks from Nebraska's Lieutenant Governor, Rick Sheehy. Participants included Northrop Grumman, Stratcom, TD Ameritrade, Booz Allen Hamilton, Inter Public Group (IPG), and the FBI. BitSight featured as NSF SBIR/IA case study at MIT
Relationship Building	Continuum Worldwide; TD Ameritrade; American Bankers Association; Inter Public Group; Gallup

Where was this startup at the end of federal government funding?

BitSight received a $1M Series Seed investment and a $24M Series A investment in 2012 and 2013 respectively that permitted the acceleration of research, development, and commercialization efforts. The investors were Flybridge Capital Partners, Menlo Ventures, Commonwealth Capital Ventures, and Globspan Capital Partners. This permitted the acceleration of the research and development and the successful launch of the company's first commercial security ratings offering. Two patents US13/240,572 Information Technology Security Assessment System and US14/021,585 Security Risk Management were granted.

At the start of federal funding there were two employees, and at the end of this funding there were fifty employees. Revenues totaled approximately $1M. As of June 2014, BitSight has three products in the market – all three leverage the core BitSight security ratings platform and are delivered via a web portal. Faced with a constant stream of evolving threats, businesses spend millions of dollars annually on people, processes, and technologies to protect themselves against cyber-risk. However, they have little visibility into the success of these investments. Without a quantified baseline, continuous measurement, and comparative data, executives cannot measure the impact of their risk mitigation efforts or assess performance against industry peers and competitors. To proactively mitigate risk, organizations need automated tools that continuously measure and monitor security performance. BitSight's Security Ratings for Benchmarking enables organizations to quantify their cyber-risk, measure the success of their overall security program and benchmark their performance against industry peers. BitSight analyzes, rates, and monitors security.

Where's the startup now?

BitSight enables continuous security ratings that minimizes exposure to data breaches and allows faster decisions with clear visibility into cyber risks across a businesses' ecosystem. It transforms how companies manage third- and fourth-party risk, benchmark security performance, underwrite cyber insurance policies, and assess aggregate risk with security ratings. A business can scale appropriately through continuous monitoring of all third and fourth parties, make security and risk decisions with speed and effectiveness, enable collaboration across the business ecosystem through consistent, data-driven security and risk communications, reduce exposure to data breach, and deliver board-ready presentations on the cybersecurity posture of an organization and supply chain. Today, over 750 customers monitor 95,000 vendors in their supply chains.

BitSight acquired Anubis Networks whose Mail Protection Service (MPS) is a secure email service that protects against ransomware, spam, business email compromise (BEC), spoofing and phishing. MPS is based on Anubis Networks Threat Intelligence ecosystem, which enables detection and avoidance of the latest and most advanced threats. This system is able to integrate with any email system, such as Microsoft Office 365 and Google Apps for Work known as G Suite.

BitSight Technologies, Inc. has so far raised a total of $92.3M in VC funding, with a series B $25.6M investment from Comcast Ventures, and a Series C $40M investment from GGV Capital. The current number of employees in the company is approximately 210.

Case Study Questions

1. Assess commercial opportunity/potential and value proposition
2. Analyze the business strategy/business model adopted
3. Identify risk elements (technical, team, market, finance); how to manage/mitigate them?
4. Growth/exit strategy – organic growth, be acquired or IPO?
5. Barriers to entry for BitSight, and its competitors
6. Critique the role and impact of NSF SBIR, and IA
7. Need for strategic partners
8. Regulatory issues
9. Take-away(s)

13   A Spectrometer in Your Hand

Technology Sector: Instrumentation
Startup: Active Spectrum, Inc.
Website: www.activespectrum.com
Location: Foster City, California

Federal Funding Timeframe: July 2009 – September 2015
Funding Amount: $1.24M

This narrative is a story of: Technology pivot(s); multiple grants; IP play; bootstrap growth; founders/team issues; M&A – acquired by Bruker Corporation

Startup Formation

Two brothers, James, and Christopher White started Active Spectrum in 2004 in Cambridge, Massachusetts. When asked what motivated him to start a company, James said - "Ever since I was a child, 9-10 years old, I wanted to have my own business. I was born wanting to do certain things. My parents were always encouraging. In fact, I wrote in my graduate school application to MIT that I wanted to start a company based on my doctoral dissertation."

In ninth grade, James with a friend set up a bulletin board system (this was before the internet). They got an account with a wholesale company selling computers, hardware/software accessories, and other components. They sold items from that company's catalog through their bulletin board. His mother would drive them to deliver these packages to customers. Although margins were thin, they continued this business until they graduated from high school. James believed that his purpose in life was to start a company and that he always had an entrepreneurial bent. He has never had a job that required working for another person.

Any one at MIT you looked up to?

"A fellow student became Vice-President at a large firm quite rapidly. He later started many companies, one of which was very successful. Then there was Alex Slocum, my Ph.D. advisor, who built a company in the electronics space that was acquired in the 1990s. He then started a rapid prototyping firm that grew to a $100M company that is still in operation. He had startup credibility, gave us space and resources, introduced me to people – all the good things a thesis advisor would do."

While he pursued a postdoctoral position at the National Institutes of Health (NIH), James continued his affiliation with MIT working closely with Alex for about three months developing a business plan. A friend, who came up with the name 'Active Spectrum' for the company, joined to help. In 2004, this friend got an offer from pre-IPO Google and left to join that company. Meanwhile his brother Chris who had just then completed his Ph.D. from California Institute of Technology ('Caltech') joined James. Because MIT startups were in general heavily oriented towards venture capital (VC), they crafted a business plan that would interest venture capitalists. They were finalists in the MIT $50K business plan competition – the prize package included the services of a law firm that incorporated the company. The two founders unsuccessfully spent over a year trying to raise VC money.

Technology

Their business plan involving technology based on radio frequency components for cellular phone base stations was reviewed by practically every VC firm in Boston. Chris even offered a microwave filter component for high-performance radios that he had developed in his Ph.D. work. A VC suggested they try the Small Business Innovation Research (SBIR) program. They succeeded in obtaining a series of contracts from the Air Force SBIR program. These helped to further develop the technology but after a few years the Air Force cancelled their multibillion-dollar advanced radio program. It is a risk you take, James notes, when the government is your only customer. By now the company had relocated to Silicon Valley. A microfluidic valve system that James had previously developed became the basis for another Department of Defense contract to make a primitive sensor to measure air particulates using extremely small signal signatures. Later this concept morphed into the development of a soot sensor[19] for which the company was awarded a National Science Foundation (NSF) SBIR award. In 2007, the founders recruited the company's first employee. It soon became clear that this sensor could be used to analyze properties of oil and other viscous liquids. The team cobbled together a rudimentary sensor package and began to market it to oil companies. In 2008, the company had its first commercial sale. This first-generation product could analyze the properties of crude oil catalysts and additives. In 2014, their fourth-generation product successfully concluded a field trial with their first paying customer in China and is now being used commercially. This strategic petrochemical industry partner has a thousand such customers around the world.

Can you describe your intellectual property strategy?

"We have an extensive IP portfolio that includes some worldwide patents on our technology. The value of our IP will not become clear until a viable the commercial model is established. This spectrometer technology development took over eight years – can you imagine, we started with radio frequency components, then microwave filters, microfluidics, and finally miniature spectrometers. We sold our first sensor package for clients' internal test and evaluation in late 2008. It was only in 2011 that we had a stable, sturdy, and robust sensor that was sensitive enough to be a commercially valuable product. It took two more years to even begin the very first field trial, which was focused purely on safety considerations. Now, finally, in 2015 we are selling these systems worldwide. We've taken an approach of continuous product improvement which has contributed greatly to the breadth of our IP portfolio. The original low-cost soot sensor was meant for the auto industry. Driven by standards and regulations, millions of cars would have required it, but no one bought these in sizable quantities. Only Caterpillar bought a single sensor package - "One cannot fight the market. We had a rough patch in 2009 during the recession; then funding kicked in – some money from my old postdoc program with NIH. In 2010, we developed a micro-spectrometer for academic researchers, a low-performance, low-cost version of what we sell today. Sales of this instrument were then our main revenue source; there was a small warranty component, but it was not a major cost. Lots of proposal writing and efforts at raising money; Caltech facilities came in handy – we were allowed access to a multimillion-dollar lab in the campus. After we got some significant research funding in 2006, an angel investor provided $150K to buy equipment, and we used heavily discounted pricing provided by Air Force contracts to buy the equipment we needed."

It was good fortune when James presented at a conference that a senior oil industry scientist was in attendance - he had the ability to connect disparate pieces of technology for useful applications in the oil and gas industry. Regarding key technological milestones, it was in 2008 that the company had its first inkling of commercial traction. Air Force contracts for development of the radio frequency component did help sustain Active Spectrum but the Department of Defense never bought anything. Their second-generation instrument was an order of magnitude more sensitive. The company started to get more orders from academics and oil companies. Development funds from a strategic industrial partner provided them a clear set of requirements orchestrated by that same scientist from behind the scenes. It was four years after the conference that James finally got to meet their internal champion in person. The proprietary micro-electron spin resonance (micro-ESR) sensor technology allowed Active Spectrum to reduce spectrometer size to one that you could hold in the palm of your hand when before such an instrument package would take up a room. Miniaturization and massive cost reductions in turn opened entirely new applications in the oil and petrochemicals industry.

Customer Discovery and Business Model

It is prohibited to use NSF SBIR funds for business development.

James insists that for instrument sales it is critical to advertise. A white paper in a high-profile trade journal and a paper presented at a conference directly led to the sale of six systems. A trade magazine such as the American Laboratory is read by tens of thousands of people in the scientific instrument's community. James created a professional website and marketing materials for a few thousand dollars, and paid for keyword search, remarketing, targeted banner ads, and print ads. He buys mailing lists to send brochures and product specs, regularly attends trade shows, and sets up a functional booth in the ones that could lead to sales. The company can respond to a few hundred quotes a year. To increase the number of quotes, he does ad campaigns around new product features, uses direct mail, internet marketing, and attends conferences. How to effectively manage leads with sales cycles anywhere from a few months to a year is a challenge for a small startup without dedicated marketing and sales teams.

Miniaturization and integration into small form factors for a low-cost relative to the market for such instruments is the value proposition. It also lies in the possible insertion of this sensing technology in applications where for a variety of reasons it has never been used before or because competing equipment is physically too large and/or will not survive in harsh environments. Active Spectrum's low-cost benchtop unit is used by chemists in industrial and university labs. The mid-size units are mostly used by academic researchers in medicine, physical sciences, and biology. The online instrument, the size of a breadbox, is used for process monitoring mostly in oil company production lines. It permits remote monitoring and control in hazardous environments. It is compact and lightweight, designed and packaged to survive dropping, heavy vibrations, and high ambient temperatures of up to 45^{0}C. The price range is between \$35K-\$75K per instrument, plus accessories, with no after-sales service or maintenance but with an available warranty and software upgrades. Initially damage or defects if any were 100% attributed to shipping issues so now the instruments are over-designed to survive shipment to countries around the world. A competitor from Belarus, whose principal business is X-ray scanners, reentered the market. This firm previously had a distribution agreement with a German company that folded and was for a time undercutting by bidding low and scooping up much business. Active Spectrum responded to this competition as most companies do, by lowering its prices and adding more features to their product.

"I've been cured of illusions and have reduced my price to make sure we have the lowest bid. In terms of risk, I never expected this new entrant. What they do now is what I did years ago but I still must compete by lowering my prices. Technological breakthroughs are another tangible risk. However, the distribution channel should not be underestimated. Industrial partners will push our product globally into entirely new markets. One partner, for example, has a 70% global market share for their core businesses."

Financing

Startup funds did not involve venture capital. Active Spectrum obtained multiple contracts from the Department of Defense SBIR program, a small amount of funding from NIH, and funding from the NSF SBIR program. The initial focus was on the automotive sector, a very challenging and difficult one for a startup. The company struggled when the Air Force terminated their multibillion-dollar acquisition program from which the company had obtained multiple contracts. Then there was the Big Recession of 2008. Active Spectrum did not seek venture capital after their unsuccessful foray in 2004-05. Their current market understanding, and commercial conditions led them to conclude that venture capital terms for this relatively illiquid instrumentation market were inappropriate. It was also felt that to yield control of the company to VC firms at this juncture was imprudent, and that their returns from an acquisition would be greater than it would be in a VC-funded scenario. The company bootstrapped its way for ten years with grants and contracts, and now shows strong sales growth.

How did you find out about the NSF SBIR program?

"In 2002, a program director from NSF that I had met suggested that I participate in a SBIR panel, and so I did. That experience was very useful, it helped me better understand the dynamics around a winning proposal. Moreover, I found out later as a grantee that NSF money, being a grant, was flexible, that unlike a contract I was allowed to respond to the market, change the product development focus. I initially treated it like a contract, only when I saw market traction, I shifted gears to pick what was in front of us."

Did this program help sustain the company?

"It would have been impossible without SBIR; we also wanted to avoid the grant mill trap; it would have made us less hungry for success. We had many ups and downs – for example, right after our very first sales, the recession hit. Most people couldn't survive. The phone didn't ring for three months; even the telemarketers had gone under! The NSF SBIR Phase I and Phase II awards made all the difference. We would not be standing today without it, along with a timely Air Force contract. An angel loaned us $150K to buy equipment with a repayment schedule and in exchange for less than 5% of equity which he still retains. Caltech too helped with lab access and space. I also had to put my own money into the company – needed capital for marketing to grow sales. At one point, I literally 'bet the farm', securing an SBA loan with my primary residence. At this point, I have no clear vision what I'd do with VC money, I don't need it and I have no interest in outside investors. We can borrow from banks on good terms."

Active Spectrum is by and large owned and controlled by James. His brother Chris owns slightly less than twenty percent. He left in 2011 and now has no active operational role in the company. James does not think there will be any need to raise capital in the next few years. He maintains that the company's suite of products is fully formed, mature and reliable, and they perform well in the market.

Company Growth

There are now eight fulltime Active Spectrum employees. Most are technical personnel including an applied chemist. All marketing and sales are done by James. He had previously hired salespeople, but it did not work out well for they lacked any real understanding of the products. James believes that to sell these kinds of instruments requires deep technical knowledge.

How do you retain people you cannot afford to lose?

"There are two key people on the team. Both have equity in the company and will share in the upside. One of them has been with me from almost the very beginning, and three others have been with the company for over three years. It is a cohesive and loyal team. All employees have enough incentive in terms of equity to continue; the vesting period is four years, 25% a year, additional options also vest in four years. The core team understands their importance. We are like family, plenty of camaraderie, everyone works hard, put in long hours, each has an office with a door; technical work requires time to think quietly, without distractions. I expect this team will hold tight. It is essential to keep everything in-house, the complexity of testing is such, and we want to make sure every product is perfect before it is shipped. The beauty of Silicon Valley is that highly skilled people are abundant here. Staffing up is not a problem and it is possible to hire very good people. For example, Varian was recently acquired by Agilent and eighteen months later that entire division was shuttered. Agilent let go close to ten thousand people working in that company's core expertise, which is instrumentation, just what we do!"

Did you make mistakes with hires?

"I made one very big mistake. The two co-founders assumed roles that were equal; Chris took care of technical aspects, ran our lab, while I took care of finances, marketing, and sales, but these responsibilities and roles were not clearly defined. Someone's got be the boss – running a company without a single boss leads to lots of problems. One person must have the controlling vote and not exercise it gratuitously. To have a co-founder who is also a family member exacerbates everything. With equal ownership, there is no deciding vote, and it becomes subtle, tricky, and sometimes underhanded, mostly the impact is negative. Anyway, finally we somehow resolved it - Chris retains equity, he earned it, he showed the technical way forward. His stake is less now because he didn't participate in the later investment rounds."

Who does the negotiating and deal-closing?

"Most contract negotiations are handled by me unless it is a multi-million, multi-year contract. I use legal counsel only when it matters, and only at the very end. Most lawyers don't understand contracts, and those who do, charge a lot. Right now, I handle business development, all by myself. Yes, this is a weakness, a one-man business development team! I really need a capable salesperson to rigorously run down all these sales calls. Marketing and sales are overwhelming – messaging, finding opportunities and leads, responding to leads."

The revenue stream, currently robust, consists of sales of three variants of the company's core product. The firm will ramp up production and sales depending on how their industrial partnerships transpire. Active Spectrum is vigorously scouting the market for new opportunities and plans to increase the 'ad spend' to engage several different market channels. The company intends to grow by feeding the pipeline with a goal of five hundred bids a year amounting to approximately $5M in sales with a 35% hit rate, the company's historical average. This bid volume will require hiring additional staff. Regarding the future, an acquisition is the preferred exit strategy, if the potential acquirer is willing to place a fair valuation. If not, the business is self-sustaining and there is no pressure from outside investors for an exit.

Key Relationships

"We view our industrial partners as key relationships. There has been a lot of consolidation in many industries with partners being acquired and merged. This means having to deal with a constantly changing management landscape that can often derail negotiations midstream."

What if a key connection sours?

"It can be a big problem although we do have a whole other line of business consisting of academic researchers and industrial chemists from where we can derive growth and profits. Diversification is the key; instruments for research purposes are one-off sales whereas enterprise-level partners offer the best potential access to new markets. In the middle is methods-development that stipulates a certain set of tests such as in food safety. This calls for third-party labs performing a series of tests using our instruments. Then there are solicitations for equipment that involve competitive bidding. But these are all unlike an exclusive enterprise-level relationship which could be a big win."

Do you have other important relationships?

"NSF matters greatly. The Phase I we got from NSF during the recession, I think it was the American Recovery and Reinvestment Act money, was a lifeline in 2009. Moreover, it was a grant, not a contract, which gave us more flexibility to align with the market. We are exploring other opportunities, like a medical device application for radiation measurement and injury. There's interest in this from the federal government involving new rules and regulations, and measurement of radiation for large civilian populations."

The Startup Experience

As a child entrepreneur, and now with ten years or more as a founder of Active Spectrum is there anything you would have done differently?

"If I had to start from scratch, I would take great pains, take great care to find an innovation that pertains to or solves a problem in a larger and more liquid market, to be able to develop technology and sell and scale more quickly to bigger volumes. This is difficult to do in scientific instrumentation. It's very hard to get liquidity in instruments technology; before I had no intuitive understanding of liquidity and inelastic demand, which is the case with business-to-business (B2B) capital equipment. There are too many competitors chasing too little demand. So, we must create demand, which is hard work."

"I remember at a national SBIR meeting in 2007, someone got up on stage and said that it took him 12 years to reach $85M in company valuation and I said to myself, well, I'll do it in 5, that 12 years was too long. Now I know better; that high-tech electronic instrumentation takes too long; people should be prepared for a long road. Therefore, two things – a market-based approach for a large liquid market, so you scale-up fast. And generate momentum; the second would be to source technology more widely, not just what I invented myself, otherwise the technology development, acquiring the technical knowhow is a hard climb, it wears you out. Be able to hire the people you need, not just technical types but for sales and business development too."

If you make serious money in an exit, what would you do next?

"I'd probably do another startup keeping in mind the points I just made, about liquid markets, sourcing technology, et cetera. I'd like also to give back, donate to further personal interests, mentor other startups; advise them on things I have learned the hard way."

Does luck have anything to do with success?

"Luck – it has helped tremendously, second only to working long hours and being able to successfully raise funds. It was serendipity, blind luck that when I was giving a talk at Rice University that a senior scientist from a company that has become a major partner was in the audience. Just that one chance encounter has led, so many years after, to a major new market for us. Luck is the underrated facet of most business success stories; it is a conceit that it all happened because of you and your skills alone."

Can you describe a high- and low-point in your startup experience?

"A low-point would surely be the death-fight with my cofounder for almost a year in 2011, and that too with my own brother. We almost voluntarily shut the company down. Someone has always to be in-charge; a wrong cofounder can be a big problem."

"One high point was when it became clear that our technology really worked and worked in the field! Our technology was validated in the market! Moving forward, an acquisition will be the end high point. For me it would be an outstanding success, after all this hard work and time spent. I'll derive great personal satisfaction."

How did your personal qualities help?

"A high level of tenacity and determination; I must be unreasonably optimistic; I cannot quit, I couldn't conceive of doing anything else. If one has a grounded understanding of reality, then you'd never start. Sometimes you must indulge in a sort of irrational thinking."

Do you keep in touch with Alex Slocum?

"Absolutely! Alex is a good friend, we go biking, skiing, and next year we are going up to the summit of Mount Kilimanjaro 19,000 feet high; he continues to do research and teach at MIT."

Where was this startup at the end of federal government funding?

The company was awarded a supplemental award of $0.5M with which it developed a higher performance version of their core technology and adapted it to serve the research and petrochemicals markets which had emerged as their two key applications of the technology. This has resulted in higher commercial sales and wider adoption of the technology worldwide. Their main product line which includes the Micro-ESR Benchtop

Spectrometer, the Extended-Range Micro-ESR Spectrometer and the On-line Micro-ESR spectrometer are directly derived from the NSF-funded SBIR project. Commercial sales of this product now total $2.01M and are ongoing. At the start of SBIR Grant (FY 2014) the number of employees was five with total revenue of $1.05M, and at the end of the grant the number of employees was seven with total revenues of $1.43M.

Active Spectrum commercialized its technology through a variety of marketing channels. The company made extensive use of digital marketing via their website, banner ads, display ads and direct email channels. It has found direct email to be a valuable channel to generate new leads. The company supplements its digital marketing efforts with frequent attendance at trade shows targeting their core customer base, which are academics, and petroleum and petrochemicals scientists. It also buys print ads in specific journals that target their niche markets. In 2015 the team was on track to generate approximately 250 sales leads through all channels combined, which represents 200% growth in sales leads compared to 2014. Historically the company has converted 35% of incoming leads into sales, therefore based on growth in incoming leads and its historical performance, it projects significant sales growth in the coming years.

Where's the startup now?

Bruker announced that it has acquired Active Spectrum Inc. in Foster City, California, the inventor, and manufacturer of innovative benchtop micro-ESR spectroscopy systems used for chemistry teaching and research, as well as for applied and industrial applications. Financial terms of the acquisition were not disclosed. The compact and fully integrated micro-ESR systems developed by Active Spectrum expand Bruker's existing research and high-end ESR product line with unique, robust, and easy-to-use benchtop discrete sample or on-line ESR systems, with remarkably good performance. Active Spectrum's micro-ESR systems will be integrated into Bruker's ESR/EPR (Electron Paramagnetic Resonance) portfolio and distributed via Bruker's global sales, service, and channel partner network into applied, industrial and education markets.

ESR is used for the detection of paramagnetic species, and its applications include the analysis of petroleum and lubricants, industrial sensing apps that include the detection of reactive oxygen and nitrogen species. In addition, micro-ESRs are excellent teaching tools and can be used for undergraduate chemistry research.

Dr. Iris Mangelschots, President of the Bruker BioSpin Applied, Industrial & Clinical Division, stated: "With this acquisition, Bruker has strengthened its expertise and leadership in benchtop and on-line ESR. Active Spectrum's unique, performance leading micro-ESR systems offer an outstanding platform to develop tailored solutions for the industrial and applied markets using very compact magnetic resonance technology."

James R. White, Ph.D., President, and Co-Founder of Active Spectrum commented: "I am delighted that Bruker, a leader in magnetic resonance known for innovation and dedication to excellence, has acquired our micro-ESR business. We are thrilled that our novel and innovative approach to this technology has been recognized and adopted by Bruker. The addition of our micro-ESR systems to their portfolio will give our customers a wide range of options, backed by Bruker's global support and service. We are excited to continue to develop new micro-ESR applications for robust, effective and affordable benchtop solutions for applied and industrial markets."

Since its founding in 2005, Active Spectrum has pioneered miniaturized ESR technology for the measurement of free radicals. Applications include analysis of petrochemicals additives, crude oil, on-line monitoring of lubricants, the shelf life of food products and biomedical research. Active Spectrum's products are used worldwide by both industry and academia.

For more than fifty years, Bruker has enabled scientists to make breakthrough discoveries and develop new applications that improve the quality of human life. Bruker's high-performance scientific research instruments and high-value analytical solutions enable scientists to explore life and materials at molecular, cellular, and microscopic levels. In close cooperation with our customers, Bruker is enabling innovation, productivity, and customer success in life science molecular research, in applied and pharma applications, in microscopy, nano-analysis and industrial applications, as well as in cell biology, preclinical imaging, clinical research, microbiology and molecular diagnostics.

Case Study Questions

1. Discuss the choice of technology – illiquid market; long road to success/failure
2. The idea of creating demand – e.g., scientific instrumentation market
3. Team building - more business development personnel; importance of technical sales/marketing
4. Finding a strategic industrial partner(s); partner relations and aligning expectations
5. Comment on the organic growth business model
6. The question of scale and inelastic demand in B2B capital equipment sale
7. Discuss the issue of more widely sourcing technology

8. The role of the federal government in long technology development cycles
9. Diversifying/additional adjacent markets
10. Take-away(s)

14 Chips Unable to Keep Time

Technology Sector:	Semiconductors
Startup:	Blendics, Inc.
Website:	http://blendics.com
Location:	St. Louis, Missouri
Federal Funding Timeframe:	January 2008 – August 2014
Funding Amount:	$1.26M
This narrative is a story of:	The need for private equity funds; bootstrap growth; extremely competitive market; serial entrepreneur

Company and Team

Blendics, founded in 2007, is based in Bridgeton, Missouri. It is an Electronic Design Automation (EDA) startup formed to provide design tools and services to companies developing complex, proprietary, low-power integrated circuits (IC). The company's products are based on research performed by the founders at Washington University and Southern Illinois University.

The IC designer's challenge is to deliver greater speed and lower power consumption in electronic devices that work reliably and consistently under increased processing complexity. This is achieved by optimizing communications between the components of integrated circuits. Traditionally, these components communicate synchronously via a global clock that controls each individual component and forces them to talk to each other in lockstep. Within each IC's design there are defined communication pathways between components ranging from point-to-point, one-to-many and many-to-one connections, some unidirectional and others bi-directional. In current designs, the global clock tree is a structure with extremely stringent timing requirements that are difficult to manage. In addition, the clock tree takes up increasing amounts of space and consumes increasingly more power.

As communication speed increases and the paths between components get longer, it becomes physically impossible to meet clock-imposed timing requirements without adding timing buffers. These however add several unwanted effects such as increased latency, decreased performance, increased power consumption and increased overall design complexity. Many components operate at different speeds necessitating splitting into multiple clock domains, each requiring one or more clock domain crossings, at the heart of which is a synchronizer that makes sure data accurately flows from one clock domain to another. Power constraints in these synchronizers make it increasingly difficult to reliably determine which state a synchronizer is in. This makes it more difficult for synchronizers to perform their function resulting in increasing risk of failure due to a phenomenon termed metastability.

It is therefore necessary to find a better way to support more robust communication requirements while not requiring design teams to discard existing intellectual property (IP)-cores or having to learn new approaches to designing IP-cores. Blendics' answer is a globally asynchronous design methodology which breaks the IC design into small independently operating IP-cores and then re-connecting each of the cores allowing each to communicate on its own timescale. Research on asynchronous computing done decades earlier toward the goal of arbitrarily large, discrete-component computer systems thus becomes relevant again, this time at the microscopic scale. These older clockless techniques could be blended with modern clocked methods to solve the anticipated complexity and reliability challenges to achieve continued Moore's Law scalability.

The four founders of Blendics are:

Jerry Cox, CEO, Professor in Computer Science and Engineering at Washington University. He received his D.Sc. in Electrical Engineering (EE) from MIT. In 1997 Jerry started Growth Networks, a venture-funded chip-set company that was sold in 2000 to Cisco for $355M. This purchase eventually led to the top-of-the-line Cisco router, the CRS-1. Jerry has published over 100 scientific papers and book chapters and has eight patents.

George Engel, Vice-President, Professor of EE at Southern Illinois University. He obtained his D.Sc. in EE from Washington University. He co-invented many products such as Magtek's Magneprint technology and the AUDIOscreener, a portable instrument to screen infants with hearing impairments. In 1991, he co-founded LoCAT, a company specializing in computer-assisted tracking systems.

Tom Chaney, Treasurer, was a leading member of a team that developed a set of asynchronous-computer building blocks in the early design of several full-custom ICs. He was also the lead engineer for a high-speed ATM network switch, an ancestor of the Cisco CRS-1. He has authored several historic papers on metastability and holds one patent in this area.

David Zar is involved with engineering issues related to asynchronous circuits and helps define the company's role in the EDA industry. He previously designed an application-specific integrated circuit

(ASIC) standard-cell library for Mentor Graphics, a leader in EDA.

The Board of Directors consists of:

Richard Catizone, President, N-Rock Technology Consulting, a firm that specializes in the integration of people, technology, and business. Richard previously spent 12 years working for Altera, the leader in innovative custom logic solutions and credited with inventing the world's first reprogrammable logic device in 1984.

Patrick Crowley, Associate Professor of Computer Science and Engineering at Washington University, has interests that span several areas at the intersection of computer architecture and networking systems, and in the design of multicore processors and memory systems.

Greg Sullivan, CEO of Global Velocity, a developer of information-centric security solutions. Before Global Velocity, Greg was Founder and CEO of technology-consulting firm G.A. Sullivan which was acquired by Avanade in 2003 after more than 20 years of award-winning growth and with offices in St. Louis, London, and the Netherlands.

Ron Indeck, Founder and CTO of Exegy, a provider of hardware-accelerated computing appliances that process market data for U.S. and international financial organizations.

Jean Roberson, CFO of Appistry, brings almost 20 years of experience across all areas of finance.

Don Winter, Vice-President of Engineering, and Information Technology at Boeing Phantom Works, where he was one of the founders of the Bold Stroke R&D Initiative focused on advancing the state-of-the-art in cyber-physical systems.

Curtis Davis, Vice-President of Engineering and COO at MulticoreWare Inc, a developer of software designed to take advantage of parallelism in multicore processors.

Market Opportunity

Following Moore's Law, the feature size of integrated circuits continues to shrink - current IC designs have reached the seven billion transistor mark. As the semiconductor market moves forward from 45 nanometers (nm) to 28 nm and 22 nm products over the next few years, the opportunities for system-on-chip (SoC) designs will grow dramatically. Design, non-recurring engineering (NRE), and global verification costs of an ambitious new SoC product at the 28-nm scale are now approaching $100M, prohibitive for all but the highest volume products. Costs per design-start based on traditional methods will grow equally dramatically, by almost a factor of three as the market moves from 45 nm to 22 nm. Blendics' semiconductor design methodology can significantly decrease the difficulty of designing billion-transistor integrated circuits for SoCs of the future.

The company addresses an industrial problem of growing importance - global timing in large integrated circuits. Timing closure has always been a crucial step in SoC design, but due to the steadily increasing demand for higher performance ICs and the greater impact of interconnection on timing, it is becoming one of the most difficult and time-consuming tasks to complete. New techniques to gate clocks and shut down selected areas of a chip to achieve lower power consumption exacerbate the timing issues with large SoC designs. The rapid increase in the use of third-party subsystems and cores also introduce their own set of complex timing constraints. These factors all add up to a yet unfulfilled market demand: reducing the cost of timing closure and verification.

The NRE cost per design-start also grows with chip density, and the company's technology will help to expand the creation of new SoC designs. If successful, the tools being developed will significantly reduce cost, power consumption and design cycles. SoC design services using a Globally Asynchronous Locally Synchronous (GALS) design methodology for reliable synchronizer designs are planned. GALS is recognized as a promising and necessary technology as device counts increase to a billion and node dimensions decrease into deep sub-micron levels. Also, conversion of board-level products to SoCs can reduce size and power requirements and increase reliability.

In 2008, the IC design industry witnessed about fourteen thousand reconfigurable and custom silicon design starts. A typical client of an ASIC design service expects a new SoC design to have NRE costs of $25M at a feature size of 90 nm and $65M at 45 nm. A semiconductor company with a digital system in production that desires to cost-reduce that product by developing a SoC version often finds NRE costs so large that cost-reduction cannot be achieved at anticipated volumes. Blendics overcomes this barrier by providing the necessary tools and inexpensive design services to clients that already have a proven design that needs to be integrated into a SoC. It offers a unique design approach that integrates tested cores, scales up easily and minimizes power consumption in the resulting SoC products.

In the early days, Blendics' credibility will have to be carefully built on the successful completion of SoC designs for clients who are already familiar with the company's products. Customers range from modest end-equipment companies to large SoC manufacturers. Medical, energy conservation, and factory automation companies are developing products that are gravitating towards a higher level of integration and would adopt the Blendics Asynchronous Network-on-Chip (ANoC) methodology due to its simplicity and ability to quickly produce robust designs that function properly even at voltage, temperature, and process extremes. While low-volume SoC clients represent only about one-eighth of the $4.4B digital logic design services market, they are nevertheless a valuable initial target since success would lead to entry into the broad SoC design services sector.

Rapidly increasing problems with global timing closure and clock domain crossing verification is a trend that will accelerate with each doubling of transistor density. As a result, asynchronous methods will become more prevalent throughout leading-edge SoC designs. The Blendics methodology will make the transition from globally clocked to GALS systems much easier and more natural because, in contrast to present GALS approaches, their methodology pre-solves all the arcane details of asynchronous circuit design.

Technology and Product

At 45 nanometers (nm) and below, new problems in semiconductor physics, lithography, and testing appear because of process variability, both between and within chips, a major hurdle for the development of new System-on-Chip products. In semiconductor design, scalability and reliability are critical for success. Scalability allows semiconductor production with significantly more transistors every two years and allows engineers to manage the complexity associated with these increasingly dense devices[20]. Reliability makes chip failures exceedingly rare despite steady growth in their complexity. Synchronizer failure in modern safety-critical chips occurs randomly and can lead to extreme consequences. Such semiconductors are often subject to high heat, extreme cold or fluctuations in power which can expose them to operating regions beyond their design limit raising the probability of failure.

A major goal of Blendics is to provide the EDA sector with a scalable approach to semiconductor design by the increasing use of standard IC design components (silicon-IP cores, a reusable unit of logic, cell, or integrated circuit layout design that is the intellectual property of one party) as opposed to reinventing components with each new design. This reuse of components will provide large savings in the design and verification effort. The key to utilizing these design components within a new IC product is reuse, ease of integration and component independence. This last factor requires independence of clock domains among the many synchronous silicon-IP cores making up the SoC and necessitating an asynchronous interconnection network. With support from the National Science Foundation (NSF) Small Business Innovation Research (SBIR) program, Blendics has accomplished proofs-of-concept for the ANoC technology and the Delay-tolerant Asynchronous Network

Interface (DANI) that makes possible hazard-free interconnections between traditionally clocked subsystems.

Any solution to the process variability problem should address both the application-specific integrated circuit (ASIC) and the field programmable gate array (FPGA) markets. To become credible in the EDA industry, this new paradigm for SoC design must be proven to work in customer applications. FPGA devices are the quickest route to such customer demonstrations. The company has successfully demonstrated this innovative, high-performance ANoC on an Altera FPGA confirming speed advantages and reductions in design effort. The Blendics design flow is important because without a major reduction in time spent on global verification the benefits of higher levels of integration, including reductions in power and increases in reliability, will not become available. Applying this paradigm to the design of ASICs provides the largest market opportunity but the ANoC methodology must be ported to the ASIC tool flow from the FPGA flow. The company's design flow has three advantages – reduced time-to-market, reduced power with increased reliability, and reduced NRE costs.

Xilinx, a supplier of programmable logic devices, invented the FPGA, and owns half the market. Unfortunately, available Xilinx tools do not allow the simple porting of Blendics' Altera results to Xilinx FPGAs. The company has created two Network-on-Chip designs in collaboration with Exegy and Global Velocity. Exegy uses Xilinx FPGAs in their products and wishes to evaluate the effectiveness of the Blendics' ANoC-based design for their future products. Eventually a test-chip design must be developed in collaboration with an ASIC SoC partner so that Blendics' test results are compelling and support the anticipated effectiveness of their design flow. It is imperative that anticipated problems be identified and solved in the test chip rather than in a much more expensive product chip for which re-spins could be excessively costly.

Blendics provides the first commercially available toolset that allows IC designers to simulate synchronizer behavior and to identify issues before fabrication. The firm offers custom licensed silicon-IP cores that implement its patented DANI technology. This can be viewed as a wrapper that seamlessly turns a standard clocked synchronously designed core into one that is externally asynchronous. It is then ready to be assembled by ClosureACE, its design tool that provides the interconnection between DANI-wrapped IP cores creating a fully functional ANoC. This tool also supports unlimited scaling of system size and guarantees all global timing constraints are met regardless of placement of the individual cores. It frees the designer from global timing concerns. ClosureACE integrates and interconnects all silicon-IP cores of a design into a robust SoC by enhancing traditional EDA tools already familiar to developers. It will help initial target clients with low-volume needs proceed from board-level products to SoC solutions at NRE costs that fit within their budget constraints.

Another Blendics tool, MetaACE, analyzes synchronizers to verify that their reliability is adequate for the application regardless of the process used to create it or the temperature and voltage at which the IC is operated. As demand increases for safety critical SoC devices, such as those required for the cars of the future, the level of reliability required will increase by several orders of magnitude. HardwareACE is a hardware acceleration solution that improves system performance through the enhancement of existing hardware and software assets. The company identifies bottlenecks in computer processing, then designs and develops custom firmware modules for the targeted functions. A custom function library seamlessly integrates the modules with the client's current software system through a common application programming interface (API). Gains are extremely significant and can range from 150-500 times normal performance.

The firm plans to compete by offering significant cost savings over in-house design capabilities, and reduced time-to-market. The former is always a harder sell, and the latter is often a critical issue in many engineering development projects. The proposed GALS EDA tool would likely prove useful to designers using large-scale FPGAs. A competitive threat is that providers of GALS tools for ASICs could modify their product for use with FPGAs prior to Blendics' tool getting to market. Four competing companies are identified in the NoC market sector, only one of which provides an ANoC but requires re-layout of third party-IP cores, a significant added engineering expense. The three others produce NoC technology in the form of silicon-IP, but their approach is not GALS. The closest competitor is Elastix, a Barcelona-Silicon Valley startup providing elastic clocks that allow cores to run as fast as the silicon allows. Although Elastix supports an ANoC, their technology requires that all synchronous IP cores be verified anew. Although automated support is provided, this is a significant additional design effort. Several firms provide fully asynchronous systems that do not take advantage of synchronous technology despite its well-established achievements and familiarity to designers.

Business Model and Execution

The first part of the firm's business model consists of design services wherein customers engage the company to create FPGA designs with high levels of adaptability and performance. This allows customers to implement a high-performance FPGA System-on-Chip (SoC) that can be modified easily. Design effort is minimized, and modifications facilitated by a flow that partitions components, reuses tested IP cores and interconnects components by means of a flexible ANoC. Two such designs are under development - an FPGA useful in high-frequency financial trading, and another capable of carrying out deep-packet inspection. The second part of the business model involves customers purchasing seat licenses to MetaACE and ClosureACE, to reduce product development costs and enhance reliability. The latter tool will have both FPGA and ASIC versions. In the third silicon-IP part, DANI wrappers are created that make a SoC's synchronous cores appear to be asynchronous thus improving energy efficiency. There presently is a prototype DANI generator that simplifies this task, but it needs to be tested over a wide range of chip architectures, first on FPGA designs and then on ASIC designs.

The sales channel for FPGA DANI wrappers will be the reconfigurable hardware vendors, Altera and Xilinx, while the sales channel for ASIC DANI wrappers will be third-party, silicon-IP core vendors. Pricing is based on a combination of NRE, subscription, and royalty fees. The pricing of DANI licenses will be more complex because different customers will wish to negotiate different arrangements according to their product's volume and sales profile. Consequently, prices will differ for each design, based on product design complexities and specific requirements. Design-services engagements are expected to predominate early while the scaling up of revenue occurs later through ANoC engagements. The company anticipates a growing revenue stream from the sale of licenses for its design tools to be significantly augmented in later years by royalties from an expanding number of DANI licenses.

The total NSF SBIR funding to Blendics in the period 2008-2014 has been approximately $1.4M. In 2011, the company received a $1M investment from the National Innovation Fund, the Innovation Accelerator's venture partner arm, to complete the design flow for both FPGA and ASIC devices. Although the company's GALS design methodology has been recognized as a valuable approach to an inevitable problem, the semiconductor industry has resisted purely asynchronous designs due to its unfamiliarity to engineers and incompatibility with existing tools and tests.

Because problems with global timing closure and clock-domain-crossing verification will accelerate with each doubling of transistor density, asynchronous methods must become more prevalent throughout leading-edge SoC designs. The Blendics methodology will make the transition from synchronously clocked to GALS systems much easier and more natural because, in contrast to present approaches, this methodology pre-solves arcane details of asynchronous circuit design. It will become increasingly valuable as SoC designs move to feature sizes of 20 nm and smaller.

Possible exits include acquisition of the company by an FPGA vendor, an ASIC foundry or an EDA tool vendor, or the company thrives on a steady income stream of design wins without being acquired, or interest from customers in one of their prototype designs justifies the development of a stand-alone product.

NSF Commercialization Assistance and Impact

IA's role with Blendics is largely encompassed by assistance with investor introductions, potential customer introductions, due diligence, and positioning of the company's technology offering. IA initially focused on trying to help the company raise capital. It engaged with or received information from angel investors, venture firms and bankers. It soon became clear that venture capital firms were not keen on investing in the EDA space. However, the company's pedigree and technical competence indicated there was likely more value to the technology than what was being recognized by the venture market.

Feedback from potential customers and obtaining a purchase order was simultaneously sought to assist not only with revenue generation but also for market validation so that potential investors would then look favorably upon the firm. Mentor Graphics was engaged to evaluate the MetaACE tool. They appreciated its technical elegance, however more importantly, the interaction indicated to IA that the silicon-IP market, with consistent growth rates, might provide the greater economic opportunity compared to the relatively flat-growth EDA industry. Furthermore silicon-IP companies have had venture backing and have been acquisition targets. ARM, a leader in microprocessor IP and the largest player in the ASIC silicon-IP market, indicated interest but preferred to wait to see more progress on the technology development front, specifically with ASICs, before taking next steps. Blendics focused their existing technology efforts towards the automated creation of DANI wrappers which could be used in the silicon-IP industry.

IA sought feedback from likely early potential DANI adopters. Two companies, Global Velocity and TranSwitch, have customers who consistently demand increased performance from products containing FPGAs and ASICs respectively. While engaging with Global Velocity, DANI technology was tested on an FPGA-based product and provided the best quantification of the technology's value proposition to date. These FPGA performance increases indicated that there would likely be similar improvements realized in the higher volume ASIC market. Blendics' success will in part be attributable to the company's early engagement with customers as well as a willingness to adapt their go-to-market strategy. While the company sought to approach the market with two EDA tools, providing a modest opportunity, it was their interaction with the market and its entrepreneurial thinking that has positioned them for a much larger opportunity, silicon-IP using DANI.

In July 2011, building on the template, lessons learned, and momentum of the "IA @ MIT" event in January 2011, IA and Stanford

University's Marie Mookini, Director of the Sloan Master's Program at the Graduate School of Business collaborated to hold the "IA @ Stanford" event. Four NSF SBIR-funded startups were selected from across the NSF portfolio with consideration given to what would resonate with Stanford graduate students. Blendics was one of four featured startups. Each company participated in a case study providing audience members insight into startup challenges as well as NSF opportunities and support. Representatives of industry, venture, and academe, from Vodafone, Intel Capital, Peter Kiewit Institute, Johnson & Johnson, EMC Ventures, Irish Innovation Center, Cisco, and IBM Venture Capital Group among many others were in attendance.

Additionally, Stanford's location in the heart of Silicon Valley provided an opportunity for IA to hold a market focus group for Blendics with key industry players. IA gathered decision-makers from Xilinx, Juniper Networks, Synopsys, Altera, Intel, Wipro, and Cisco to pose questions and express the needs of the semiconductor industry. This opportunity allowed Blendics to refine its go-to-market strategy as well as providing IA insights into semiconductor technology pain-points that might be successfully addressed by other NSF SBIR-funded companies.

Where was this startup at the end of federal government funding?

Products and Patents:

MetaACEPRO: A software analytical tool capable of determining the reliability of both single-stage and multistage synchronizers. This analysis can be carried out at multiple operating conditions for various semiconductor processes, supply voltages and operating temperatures

MetaACELTD: A node-limited version of *MetaACEPRO* available as a free download from the Blendics' website. Engineering educators, engineering students, and design engineers will find it useful in increasing their understanding of metastability and the performance of synchronous circuits. This version of the tool is limited in the number of netlist nodes it can support and handle fully extracted netlists. This "freemium" site is expected to improve understanding of metastability and of the resulting risk-to-degradation of system stability. Analytics obtained from this site will support the proposition that there is a large demand for a reliable metastability analysis tool. Blendics has entered into an agreement with semiwiki.com to be the agent for the acquisition of the MetaACE tool by an electronic design automation (EDA) tool vendor. Bolstered by the results from this site, this agreement anticipates the completion of an acquisition deal within six months.

DANI: A patent-protected intellectual property (IP) that makes synchronous semiconductor cores appear asynchronous to other system components. A version of *DANI* for FPGA projects is available today. A version applicable to ASIC projects is under development. This technology has proven to be essential in the development of a high-performance FPGA platform that is the core of new Blendics' products aimed at customers in the financial services sector.

The company has entered into a partnership agreement with ViciTek, to provide hardware acceleration to their financial marketing services. According to this agreement, substantial compensation for Blendics' engineering services will support the delivery to this partner of hardware-accelerated products, a subsystem that can be inserted at the interface between a financial trading system and the network that performs fast risk-checks on a market order prior to sending it to an exchange's matching engine. Sales of a software version are ongoing, and an FPGA version is under development. The hardware version of this product uses both *DANI* and *MetaACE* technology. The initial ViciTek product, *Interceptor*, has received an enthusiastic reception from potential customers. The launch of this technology is planned for early 2015.

At the beginning of federal funding, Blendics had three employees, and at the end of this funding it had eight employees. Total revenue was approximately $160,000.

Where's the startup now?

Blendics, Inc. offers design services, design tools, and silicon IP that enable companies to harness the speed and performance of hardware acceleration and system-on-chip integrated circuits. It provides *DANI* wrappers that provide an interface between synchronous cores and ANoCs; *ClosureACE*, a tool that facilitates the design of an asynchronous network-on-chip that connects *DANI*-wrapped objects; and *MetaACE* for simulating and validating synchronizer performance during metastability. The company offers its technology for FPGA and ASIC markets.

Blendics has signed a partnership agreement with Altera. Headquartered in Silicon Valley, Altera Corporation (NASDAQ: ALTR) is the leader in innovative custom logic solutions, and has been, ever since inventing the world's first reprogrammable logic device in 1984. Today, over 2,600 employees in 19 countries are providing even more ingenious custom logic solutions – addressing a range of concerns, from power consumption to performance to cost – for customers in a wide variety of industries, including automotive, broadcast, computer and storage, consumer, industrial, medical, military, test and measurement, wireless, and wireline. In addition to devices, Altera's comprehensive solutions portfolio contains fully integrated software development tools, versatile embedded processors, optimized intellectual property cores, reference designs examples, and a variety of development kits.

Consider the IC designer's paradox. Simultaneously, he/she must deliver increased processing complexity, greater speed and manage power consumption, and must work reliably and with consistency. These challenges are monumental. The answer lies in optimizing communications between the IC's component parts - within each IC's design there are defined communication pathways between the components, including: point-to-point, one-to-many and many-to-one connections, some unidirectional and some bi-directional paths. In current designs, these communication paths are synchronized by a global clock tree, a structure which has extremely stringent timing requirements and is increasingly difficult to manage. Add to that, the clock tree takes up increasing amounts of space and consumes more and more power. As communication speed increases and the paths between components get longer, it becomes physically impossible to meet the stringent timing requirements imposed by the clock without adding pipeline stages. However, when these timing buffers are added, several unwanted effects are introduced including increased latency, decreased performance, increased power consumption and an increased overall design complexity.

Additionally, many of the components operate at different speeds which increases the complexity, splitting it into multiple clock domains, each requiring one or more clock domain crossings. The heart of a clock domain crossing is a synchronizer, used to make sure data accurately flows from one clock domain to another. To satisfy power constraints in these synchronizers the distinction between a 0 and a 1 has become so small, it is increasingly difficult to reliably determine which state a synchronizer is in. This makes it more difficult for synchronizers to do their job resulting in increasing risk of failure due to metastability. Designers can effectively reuse or create new IP-cores providing the additional functionality demanded for next generation devices. It is also evident that IC fabricators can effectively develop next generation techniques, creating the ability to add more and more logic to a single IC. However, the problem is that when designers connect all the components together, they are too often unable to achieve reliable communication at the required speeds.

Blendics' answer is to eliminate the global clock tree, and in its place, provide a new communication path for each component. This new approach allows each component to operate independently, and at its optimal design speed, thus achieving scalability and reliability. Synchronizer failure in modern safety critical ICs can lead to extreme consequences. These failures occur randomly and leave no evidence as to the cause of the problem. In addition, many times safety critical ICs are subjected to extreme conditions such as high heat, extreme cold or fluctuations in power which can expose the IC to operating regions which raise the chances of synchronizer failure. *MetaACE* analyzes synchronizers to verify that their reliability is adequate for the application regardless of the process used to create it or the temperature and voltage at which the IC is being operated. As demand increases for safety critical SoC devices, such as those required for the cars of the future, the level of reliability required will increase by several orders of magnitude. *MetaACE* fills the emerging need for verification of synchronizer reliability.

Case Study Questions

1. Discuss the importance of the U.S. semiconductor sector, and Blendics' value proposition and product positioning
2. Comment on the go-to-market strategy and market timing issues
3. How best to fund this endeavor? Why the NSF SBIR Program?
4. At what point did the company first seek venture funding; how did the track record and technological credibility of the management team figure into the fund-raising process; how much time did it take?
5. Discuss IA's role and value of its assistance
6. Critique the business strategy/business model
7. Identify risk elements (technical, team, market, finance); how to manage/mitigate them?
8. What are the barriers to entry?
9. Discuss company growth and exit strategy
10. Conduct a valuation exercise for this startup
11. Take-away(s)

15   Sensual Touchscreens

Technology Sector:	Displays
Startup:	Tangible Haptics, LLC
Website:	www.tangiblehaptics.com
Location:	Evanston, Illinois
Federal Funding Timeframe:	July 2012 – July 2017
Funding Amount:	$1.51M
This narrative is a story of:	The need for private equity funds beyond federal funding; IP play (over thirty patents); scale-up/ production challenges; name changed to Tanvas, Inc.; changed corporate structure from LLC to Inc.

Startup Formation

Greg Topel is the prototypical engineer at heart. He started with Legos as a kid and loved to get his hands dirty taking apart gadgets. When his father brought home a personal computer, he was so curious he took the case off to study its inside. A high school mechanical drawing class solidified his drive towards making things. In this same class, he was one of the first students to use computer-aided design on an 8088-computer system. He designed a gear and gave the computer a "rotate" command and had to come back the next day to see that it had indeed rotated! Another class he enjoyed involved magnetically levitated vehicles racing along a track. He was an entrepreneur but did not know it then because there were no courses in entrepreneurship like they have now. In college, Greg and two friends came up with ideas to start a company – one was to have the scale and cold-cuts meat slicer work together. But the concept never got further than an idea on paper.

After college, Greg worked at Motorola's Infrastructure Group, then at Visualize, a small engineering firm, and was later Engineering Manager of Cardio Equipment at Life Fitness, a global leader in fitness equipment. In 2009, he became a Manager of the Technology Center at Illinois Tool Works (ITW), an American Fortune 200 company that produces a wide variety of components, equipment, and consumable systems

"I had the opportunity to grow professionally throughout my career starting with Motorola which taught me the fundamentals of engineering, then Visualize which was very much like a startup, and Life Fitness where I played a significant role in launching several major products. During my tenure at Life Fitness, I was also able to obtain an MBA from the University of Chicago in 2007, with a concentration in entrepreneurship and finance. When I joined ITW's Technology Center in 2009, I led a team of engineers that executed innovative technology programs and was involved in several startup technology evaluations which helped to plant the seed of wanting to start a company."

"In March 2011, I got connected to two professors from Northwestern University, Ed Colgate, and Mike Peshkin. They had researched haptic technology for years and were looking for someone to commercialize what they had developed. Incidentally they had both received research funding from the National Science Foundation (NSF). I spoke to friends in the investment world and got advice from my MBA professors and business colleagues; many cautionary tales, how things may not work out, also that experience even of failure is good. I had many discussions with my wife. Even though the venture would likely be a strain on the family, she simply did not want me to later regret a missed opportunity. I have to say her support is a major reason I am with Tangible Haptics today!"

Greg saw working prototypes in the professors' lab at the university and got the chance to feel live haptic effects. Impressed by the two researchers and their technology, he started to write a business plan. Greg quit ITW in June 2011. Half-a-dozen investors showed interest. The university owned six patents related to the technology. Tangible Haptics successfully negotiated with the university and obtained an exclusive license to these patents.

"Ed and Mike are both quite involved in Tangible Haptics. All three of us share a common vision for the future of haptics. Our long-term goal is to see haptics become an important part of how people interact with touch devices. Radio and TV made sound and sight happen, now we want to add a third dimension, the sense of touch, to complement sight and sound. We believe, for instance, that the use of the sense of touch will help the vision- or hearing-impaired in their ability to interact with each other. It will be transformative!"

Technology

Ed and Mike have spent their entire research career in haptics starting in robotics and then roughly a decade ago shifting to research of surface haptics. Prior to Tangible Haptics, Ed and Mike founded two other haptic-like startups. One was a product company that they sold, and the other a consulting entity. Now, in addition to their university research they are both focused on Tangible Haptics with the intent to translate their research to marketable products. The fundamental research is university- based but the company is now generating its own intellectual property. Patents have not yet been filed. Within the year, Greg plans to move the office from Evanston to downtown Chicago for a more central location to expand the team and to help speed-up progress towards a commercialized product. Greg has in mind a specific first product that has not yet been made public - a touchscreen with a haptic response in a specific market segment targeting a specific customer need. He believes that the first step is to gain a market foothold with a simple solution. Currently most haptic effects require unique components and electronics in addition to the touchscreen.

"Our solution will replace current touchscreens at lower cost, involve minimum changes in internal architecture and electronics, and can be manufactured using current production tooling. All this will be invisible to how it looks to a customer. How we do the haptic part is based on university research that has blossomed into more R&D resulting in over twenty patent applications. The innovative part is friction modulation, while most other haptic methods are vibration based; creating recognizable haptic effects must be forced-based as they are when you touch things in the real world."

In continuing its R&D efforts, Tangible Haptics has been careful not to use university facilities, especially the labs led by the two professors[21]. Greg initially worked from home and then transferred to a small office.

"We now have a full lab with a machine shop. It is essentially like being in an incubator with rent being our only cost. The key milestones on the technology development front involve electronics and manufacturing. We are close to finalizing our chipset design by leveraging our relationship with a semiconductor company because to do it ourselves will require too much capital and manpower."

Are they based in the U.S.? And manufacturing?

"Yes, mostly in the West Coast; there's no need to go overseas for the integrated circuit design; as for manufacturing lines, we plan to remain U.S.-based as much as possible but with touchscreens 90% are made in China and Korea. We have been cultivating relationships with a couple of them."

Do any standards and regulations exist for haptic-embedded consumer devices?

"In fact, people approach us about this when we show our prototypes at trade shows. It's a new frontier. Recently, a regulations writer asked us if we could help with regulatory work. For example, regulations don't exist for such devices regarding electrostatic interference issues. There are other concerns related to compliance which we are addressing based on current regulations. Additionally, I think we can address aspects related to possible new standards and future regulations for our products with software which we plan to develop in-house."

Customer Discovery and Business Model

The company discovers potential customers mostly through professional connections and networking. On the customer front, early on there was never a phone call that Greg did not take, and rarely would he turn down a meeting. Recently Tangible Haptics was accepted to participate in the Innovation Zone at the Society of Information Displays (SID) trade show. Their presence at such events is useful for it is where they meet potential customers and strategic partners, and even investors.

"There's also the media aspect, and reporters who write stories about featured technologies. This is especially true if you have a haptic-integrated product prototype to showcase possible applications. They take you more seriously when they can touch and feel such a product. Looking back, it was critical for us to be there, demonstrating our product, being engaged, and obtaining feedback. Our whole team, including Ed and Mike, was present for four days."

"You must demo your product where your potential customers are."

What insights have you gained about the haptic market?

"We constantly talk to potential customers and users. Sometimes users of consumer devices don't know what they want. Years ago, if you had asked somebody if they'd like to have access to maps on their phone, most would have thought it interesting but not necessary. Now people absolutely rely on that feature and cannot do without it. Everyone we speak with agrees haptics will happen, but the forms and applications are currently undefined. We believe haptic technology will become ubiquitous, a must-have feature for display products. The other factor, as with anything else, is cost, and we address that by combining sensing and haptic actuation all in one."

"For original equipment manufacturers (OEMs) two factors are most important – cost and differentiation. I have a good idea of the general market but what people will specifically adopt is the exciting, unknown part. We envision people creating a wide variety of applications that expand experience and add functionality. The value proposition is that touch (haptics) provides a third dimension for communicating beyond hearing and sight, providing a truly real physical interaction with a device. Consumer electronic devices have been commoditized and this can be a differentiator."

Can you explain the business model?

"Consumer electronics is an entrenched value chain – not easy for a startup to enter. We don't own the glass due to the heavy capital costs, and therefore our product will necessarily be made by a third party. Licensing and royalty fees from consumer device OEMs will generate revenues. Our initial targets remain confidential, but it is a major market, and we will be leveraging partnerships with OEMs in those spaces. The financial model was built through a combination of help from board members, insights from industry partners, and expert advice from the consumer electronics device makers to help anticipate customer needs and develop market-entry strategies. In terms of competition, there are certainly other players who have established themselves. However, most are either vibration-based or do not have the functionality we can provide. As with any consumer electronics technology, the competition is fierce, but the combination of our IP and our knowledge of haptics provides a clear advantage over other haptic solutions."

Financing

Startup funds involved investment from founders, sweat equity, bootstrapping, and making do with the minimum. As is always the case, potential investors at that early stage were extremely hesitant to invest in a hardware company. Nowadays, there are not many sources of early-stage funding for hardware startups. Hardware companies are damned either way – if you are a component company you are at the mercy of a strategic partner, and if you want to make your own product, large capital outlays and years of technology development are required. In the early days, Tangible Haptics did receive venture and partnership offers but they were mostly one-sided.

"I can tell you, without the Phase I award from the NSF Small Business Innovation Research (SBIR) program, the company would not have survived. The first year after inception was very difficult. After I quit ITW, there was an extended period without a salary. I only started paying myself after we got the NSF SBIR award in July 2012. Then everything started to jell – we secured seed investment from a major OEM in March 2013 after having engaged with them for over a year. We also sold a few prototypes. In 2013, because we received the Phase II award from NSF, we could hire three additional employees. Success up to this point would not have been possible without the NSF program."

Greg maintains that at this point the biggest risks are financial and technology adoption. He is feverishly working to close a Series A investment round. The investors are a mix of strategic partners and venture capital investors. Two supplement proposals to their current funding from the NSF SBIR program, if awarded, will total $650K. One of them involves co-development with a large company.

How did you arrive at company valuation?

"The net present value numbers are essentially hand-waving; I think it's best to use 'comparables' if available. I considered the IP we own, the level of technology and product development, the remaining technical risk, first revenues, and the number of years to breakeven. There was also the fact that a major OEM had already invested in us. The interest shown by other potential partners also factored into calculating the overall prospects for the company. For investments in hardware companies there's not too much competition among venture capitalists (VC) – there are so few of them. Hardware companies are the ugly ducklings. The Series A investment will give us sufficient runway to get to our first product offering. The VCs we are engaged with understand our company vision, understand the road is difficult and riddled with potholes. They do not intend to rush us to revenue but are rather more interested in doing something truly disruptive."

Company Growth

Presently Tangible Haptics has four fulltime employees – a software engineer, a materials scientist, an electrical engineer, and Greg, who is responsible for business development, fundraising, and financial planning. Greg was instrumental in developing the first prototype during the initial phase of the NSF SBIR award. The materials expert and the electrical engineer are key personnel for the development of the core technology while the software engineer is essential to the development of software communication protocols and applications. All four employees have an equity stake in the company, with standard vesting periods. Greg believes that building the right team is most critical for successful implementation of the next phase.

The Series A investment will allow the company to build-out the technology development team. Investors will need to accede to the development plan and the make-up of new hires. Greg thinks that he has the experience of the last four years to continue as the CEO beyond this investment round. The company plans to soon move out of their current small office in Evanston to downtown Chicago. Being in the city will help attract the right talent with commuting convenient to downtown from a fifty-mile radius. The firm is roughly twelve months away from marketing a product and is deep into software development while simultaneously working with manufacturing partners, to have a marketable product, two years from now. With the investment in place, Greg plans to triple the size of the team to twelve engineers and support staff.

"As a startup, people are by far the most important resource you have. You simply cannot make a hiring mistake - if you do it must be resolved immediately. Right now, it's a small team, and we all march to the same beat. It is a flat organization with a mix of senior and junior personnel."

"The next two years will involve product development, supplier and partnership identification, and in parallel developing the software and the software development kit (SDK). Revenue streams will consist of licensing and retainer fees; and use of our SDK and joint development agreements to help create haptic applications for partners to incorporate into their products. Beyond the next two years, we plan to be the source for haptic touchscreens to be used in a wide variety of applications."

Current Board members are Greg, Ed, Mike, and Wes Davis, previously the CEO of six technology companies. Wes successfully positioned and sold four of these six companies for a total of $276M. After the Series A investment is completed, it is expected that the makeup of the Board will change with the addition of a VC representative.

"Ed and Mike serve in an advisory capacity. They have no operational role in the company. They consider this their technology legacy. A future acquirer, should there be one, would need to ensure their intent to carry forward with the two founders' plans to integrate haptics into a variety of applications. No one in the company is interested in selling solely for monetary gain as the application of the technology is far more important."

"Overall, the opportunity for this technology to be truly transformative is far too great for short-term gains alone."

Key Relationships

Do you think it wise to depend too much on a single strategic partner?

"At some point, we must focus – one product, one partner. We must choose a product and do it well. If we find it's the wrong decision, we must fail fast and change direction with the contingencies we have in place. It's not easy for a hardware company to pivot."

"There are many big players all looking for product differentiation in a commoditized market. We are not yet ready for a trade event like the Consumer Electronics Show. The form factor must be completed - we can't expect customers to imagine things. NSF too is a key relationship – it adds credibility; people know that the science behind the product has gone through multiple rounds of technical due diligence, that our technology has been validated by technical and domain experts."

The Startup Experience

What makes you want to be an entrepreneur?

"I still remember sitting at my desk at ITW when I was contacted about being the CEO of this startup. My wife and I had just had our second child, I had an excellent job, and the simple question was why should I leave? It was something I had always wanted to do even though I know the risks and that the more likely outcome would be failure. You ask am I an entrepreneur or did I make myself an entrepreneur? Even before college, I was good at organizing people to complete a common goal. But the one main truth about being an entrepreneur is you must be willing to believe, even when everyone says it won't work, that the market is too big, and that you simply have too much stacked up against you. It can be extremely lonely, it will completely consume you as you never really turn it off, but it is also absolutely your very own which makes it all worth the struggles and hardships. So, while the skills of being an entrepreneur are certainly learned, the foundation of the person will determine if they can ride out the bad before they can begin to succeed."

How would you describe your accomplishments so far; your high- and low-points?

"NSF SBIR Phase I funding came in at just the right time – we certainly wouldn't have survived without it. The other day I was looking at the presentation slide deck we made to our OEM seed investor, describing to them what we'd do with their investment. And you know what? We did everything we had committed to completing which is very satisfying. There are several high-points – funding from NSF, partnering with powerful strategic partners, high VC interest now in our technology, they even invite others to our demo."

"Experts say that more than ninety percent of startups don't make it beyond the three-year mark. It makes me happy we are still standing, and on the cusp of closing a major investment round. In fact, 39 out of 40 startups never close a Series A round so that alone is a major accomplishment assuming everything works out as expected. The low points are also numerous. An example is when I inadvertently submitted an incomplete proposal to NSF with an erased budget. I thought I had screwed up a major opportunity, but it turned out to be an NSF system error. The many times being told no, especially when it seemed a "yes" was guaranteed. There are plenty of others but nothing out of the ordinary for a startup."

Do you have advice for hardware startups?

"You must have at least twelve months of expenses saved up, even when it looks like you will get investment out of the gate. The road is paved with many naysayers, but if you believe that the technology you are developing has real potential, that you can create value, new functionality, and experience, then you must learn to tune-out negative comments while listening to advice that can sharpen your pitch, product, and roadmaps. It's important to have realistic expectations, a real plan and schedule. Everything will of course not go according to what you have on paper, but without a roadmap you have no idea when you are off base. Overall, you've got to love what you are doing; you must believe it in your soul. After all, it will likely become such a large part of your life."

How would you define success?

"After long hours of self-reflection, I have concluded that success for me is defined by doing something that will profoundly impact the lives of others. My wife and kids are, without a doubt, the most important people in my life, so anything that takes me away from them must have significant meaning. If this startup was only about building a better mousetrap or a minor novelty item, it would make the long nights, constant traveling, and all the normal struggles of a startup impossible to justify the sacrifices my family makes. At ITW, travel was minimal, the job was flexible, and I had time to be with my kids. Now that this startup consumes a good chunk of my time and energy, I need to truly feel that what I am doing is worth the tradeoffs. Even if I were guaranteed a major positive financial outcome, and if what we are doing was meaningless, I don't think it would make up for being away from my family. Knowing I am working on a technology that will truly transform the lives of people is critical for me to continue with the chaos that comes with running a startup."

Where was this startup at the end of federal government funding?

In 2013, this startup obtained seed round of $1M from a strategic partner. This was followed by a Series A round of $4.6M in 2015 from a private equity syndicate. The commercialization activities related to this funding was the promotion of the technology through a booth at the 2017 Consumer Electronics Show (CES) in Las Vegas, Nevada, and the creation of a demonstrator kit to enable product manufacturers to evaluate the technology. The reaction from the consumer electronic industry and press alike was phenomenal. Leading up to the show, CES provides a webpage to all exhibitors for attendees to search, receive information, and reach out to customers. Of the over 2,400 exhibitors at CES, the Tanvas webpage was the twelfth-most viewed webpage. As fantastic as that preshow recognition was for this technology, the actual reception received was exponentially greater. Besides winning awards, over forty articles/ videos were created about the company.

Since CES, Tanvas has entered dozens of Non-Disclosure

Agreements (NDA) with a wide range of customers including smartphone and tablets OEMs, automotive suppliers, casino gaming machines, kiosks, and other various public displays that are customer-facing. The company is servicing this demand with technology demonstrator kits that sell for five thousand dollars each – the kits consist of the Tanvas haptic touchscreen, a Nexus 9, all the required electronics, and the SDK. The kit allows customers to socialize the technology throughout their respective companies including upper management, experience the applications Tanvas has created, and create their own haptic experiences using the SDK tool.

Pilot projects include developing haptic outputs for laptop touchpad interactions with a strategic partner to standardize the output for the Precision TouchPad (PTP) protocol, developing a haptic touchscreen application for an automobile manufacturer, public terminal haptic user interface concepts for the visually impaired, and restaurant gaming tablet application concepts.

The following patents have been granted:

1. US 2014/0306842 – Haptic display with simultaneous sensing and actuation – June 16, 2014

2. US 2016/0349880 – Devices and methods for generating haptic waveforms - Feb. 20, 2015

3. US 2016/0357342 – Electronic controller for haptics display with simultaneous sensing and actuation – Feb. 20, 2015

4. 62/342,594 – Haptic touch screen and method of operating – May 27, 2016

The company had one employee at the start of federal funding, and fourteen employees at the end of this funding. Revenues at this point total approximately $1.5M. From the very beginning of the SBIR award, the objective has always been to commercialize the product and eventually become the new form of what all touchscreens will be. Seed-round funding coupled with federal funding enabled the company to complete the core hardware development and build the first prototype that was ready to be introduced to the market as proof-of-concept. In early June 2014, the company secured a booth in the Innovation Zone at the Society of Information Displays (SID) Display Week trade show. The show provided the company a platform to introduce the technology to the public where attendees seek innovations in the touch-interface sector. The results were immensely significant, including leads on two potential customers, two potential strategic investors, plus exposure to solidify Tanvas as a viable technology company to institutional investors. The NSF SBIR has been a major enabler for this company to complete the necessary core R&D to produce a hardware product for commercial applications.

Where's the startup now?

The ongoing Series B round seeks an additional $12 million in funding. Tanvas puts touch inside the touchscreen – the company's mission is to connect people to the digital world through rich touch interactions. Despite advances in graphics and sound, today's touchscreen remains a lifeless window into the digital world. This technology adds a new dimension – the realistic sense of touch – so one can create dynamic textures that can be felt with the swipe of a finger. The technology uses electrostatics to control friction and create virtual touch. The applications for this technology are endless – feel the edges of keys, the snap of a toggle switch, the swipe of a turned page, the direction and magnitude of impacts in a game. For anyone who wants to elevate, deliver, or participate in a more engaging and complete touchscreen experience, Tanvas provides a touchable canvas on which to create. Its programmable tools allow OEMs, agencies and developers to imagine and create an infinite number of holistic and integrated experiences on any touch display.

"The major drawback of touchscreens has always been the missing element of physical feedback. We constantly interact with the world through the sense of touch. Yet until now, that has not been possible in our electronic devices. Tanvas finally empowers touchscreens with what we didn't know we were missing: true interaction with what you see."

Case Study Questions

1. Assess the commercial opportunity and value proposition
2. Barriers to entry, for both the startup and its competitors
3. Identify areas where the company needs help
4. Comment on team building and team dynamics
5. Identify risk elements; how to manage and mitigate them
6. How to properly time market-entry; develop a go-to-market strategy
7. Discuss the growth and exit strategy
8. Conduct a startup valuation exercise
9. What are the issues around transferring technology from a university?
10. Is this the right business model?
11. Take-away(s)

16 No Need for Spiders to Make Silk

Technology Sector: Synthetic Biology
Startup: Refactored Materials, Inc.
Website: www.boltthreads.com
Location: Emeryville, California

Federal Funding Timeframe: July 2010 – February 2016
Funding Amount: $1.18M

This narrative is a story of: The need for private equity funds; strategic partners; rapid growth; IP play/issues; business model changes; market pivot; scale-up/production challenges; name changed to Bolt Threads, Inc.

Company and Team

Refactored Materials ("Refactored"), founded in 2009 by three recent graduates of the University of California at San Francisco (UCSF) and the University of California at Berkeley (UCB), is a Delaware corporation specializing in synthetic biology and microfabrication to commercialize materials which are too scarce or too expensive to extract from natural sources. Specifically, the company is developing synthetic spider silk for technical, structural, and textile uses. Spider silk outperforms all available competitors on a variety of metrics but has never been successfully commercialized. The startup aims to bring together the best features of natural fibers and synthetic processing to create a high-value silk fiber for apparel designers and consumers.

Natural fibers such as cotton, wool, and silk are highly prized for their combination of tactile feel, tensile strength, breathability, comfort, and improved sustainability over petroleum and cellulosic products. They are generally harvested from agricultural processes susceptible to weather variations, disease, and pests. In addition, they show season-to-season variability in quality and offer limited ability to change material properties. While natural fibers are widely thought of as sustainable in general terms, modern high-density agricultural methods are not environmentally benign – silk, for example, consumes more than thirty thousand metric tons of water per kilogram of silk fiber produced. It is now possible to create novel fibers that provide the comfort, feel, and breathability of natural fibers combined with the precise control of properties of synthetic fibers. Spider silk is considered a marvel in the world of natural materials. It possesses high tensile strength, high extensibility, and unrivaled toughness compared to common synthetic fibers. It is also lightweight, breathable, and flexible. If an inexpensive, high-quality, reliable supply of synthetic spider silk were to exist, a wide range of industries from medical devices to sporting goods to cosmetics would benefit. The company's silk engineering platform enables just such a series of fiber and fabric products along with consistent pricing and true sustainability.

The current executive team consists of founders Dan Widmaier, Ethan Mirsky, and David Breslauer, and Rena Hill. Dan, CEO, received a Ph.D. in chemistry and chemical biology from UCSF, and specialized in synthetic biology. His research involved the design, construction, implementation, and characterization of synthetic genetic circuits to control natural microbial organelles, specifically directed at the production and secretion of spider silk protein. He previously worked at Amgen where he developed and applied novel bio-analytical methods for monitoring the scale-up manufacturing of clinical drug candidates for oncology and inflammation diseases.

Ethan, COO, completed a Ph.D. in biophysics from UCSF with a specialization in synthetic biology. His research involves the engineering of protein signaling networks, microbial metabolic engineering, spider silk protein expression, and the refactoring of bacterial organelles. In 1996, Ethan co-founded Silicon Spice, a high-performance telecommunications semiconductor startup. He was involved in all aspects of business and technical development, including inventing novel chip architectures and software systems, acquiring over $90M in seed and growth funding, and an eventual acquisition by Broadcom in 2000 for over $1B.

David, the Chief Scientific Officer, earned his Ph.D. in the joint bioengineering program at UCSF and UCB specializing in microfluidics and micro-device engineering. His research focuses on the design, manufacture, and use of microfluidic devices for biotech applications, with specific focus on the microfluidic processing of silk proteins into fibers, the rheology of silk solutions, and the computational study of biological fluid flow.

Rena Hill is an experienced scientist, lab manager and technician well-versed in the latest synthetic biology tools. She has collaborated with both Dan and Ethan on their work with spider silk genes, and therefore has unique and invaluable knowledge of the synthetic biology, expression, and purification of spider silk.

Refactored Materials draws on an extensive network of advisors and investors. A scientific advisory board includes Sam Hudson, professor of fiber science at North Carolina State University, Chris Voight, professor of biological engineering at MIT, David Kaplan, professor of biological engineering at Tufts University where he leads a NIH research center for the use of silk as a tissue engineering material, and Susan Miller, professor of chemical engineering at UCB.

The company also draws on experts in the textile industry such as Sue Levin, the founder and former CEO of Lucy Activewear, Steven Arcidiacono, a leading U.S. Army textiles researcher, and Greg Haggquist, the founder of textiles innovator Cocona. Peter Hecht, the founder, and CEO of Ironwood Pharmaceuticals, provides guidance on company management, and building for success. Investors Steve Vassallo, a general partner at Foundation Capital, a venture capital group located in Menlo Park, California, and Paul Koontz, another general partner in the same firm also provides advice and oversight. Michael Shuster from Fenwick and West is the company's IP counsel and Lynda Twomey the corporate counsel.

Market Opportunity

The three-trillion-dollar global textile and apparel industry is a voracious consumer of natural and synthetic fibers used in the production of products ranging from apparel and upholstery to sportswear and high-fashion. The growth in worldwide textiles demand is buttressed by growth in population and average income. Access to novel textile sources drives the designer's ability to brand and sell differentiated products to consumers. Nearly all cultures worldwide view apparel selection as a statement of self and a reflection of personality. Many consumers assign high-value premiums to attributes such as sustainability and ethical animal treatment in the production of textiles and apparel. This combines with traditional factors such as price, feel, and trends to form a broad spectrum of differentiation between natural fibers and synthetic petroleum-based fibers, leaving a market ripe for a new class of fiber that incorporates the best aspects of both.

In contrast to natural fibers, nearly a century of chemical engineering experience has culminated in the development of a robust variety of petroleum- and cellulose-based polymers made into an array of fiber types. These fibers are produced using efficient chemistries and spinning and can be modified to meet customer demand for tensile strength, dye retention, and other desirable features. Synthetic fibers are an inexpensive alternative to natural fibers, although they remain inextricably tied to the price fluctuations of petroleum and cellulosic feedstock. These fibers lack durability in the case of cellulosic fibers, and other innate features such as breathability, drape, and feel compared to their natural counterparts. They are also considered unsustainable. It is also true that the fiber industry has seen only modest product innovation in several decades.

Sericulture is practiced in over thirty countries worldwide but approximately 90% of the production takes place in China and India due to the specific climate required for raising mulberry bushes, and access to inexpensive manual labor. This industry is composed of specialists who raise silkworms and harvest silk fibers from the cocoon. It is currently in flux with rising wage pressures, and silk cocoon production gaining traction in other Southeast Asian countries and Africa. Silk production is a complex and labor-intensive process. The leaves from mulberry bushes are the only source of food for domesticated silkworms. The worms are fed leaves in a high-density format until they make cocoons that protect them while they undergo metamorphosis to adult moths. Cocoons are harvested within forty-eight hours of construction completion with the pupae killed with steam. They are dried, sorted, and graded by hand, and then passed to "reelers" who boil the cocoons in a mild base to remove the protein glue that surrounds the silk fibers. Workers unwind the cocoon by finding fiber ends and attaching five-ten fibers on a reeling/twisting device to make raw silk yarns. These yarns are then graded and sold as the base commodity and as input material for knitters and weavers to make fabrics.

Textiles are an ideal first market for Refactored due to its massive size and ease of entry. It also does not require the highest-end material properties. This allows the company to rapidly achieve product-quality material and revenue streams. Prospective customers include design houses of the most elite haute couture, ready-to-wear designers, department stores with their own lines, and mass market retailers. Silk textiles are often made to a designer's specification. Fabric is dyed and finished and sent to a garment factory for cutting and sewing. Finished garments are then delivered to the designers' distribution chain for sale to consumers.

The startup's business model is to sell its product to designers as an alternative, at a competitive price, while retaining the beneficial properties of sericulture-derived silk. The advantage is that automated fermentation and mechanized spinning eliminate the high-labor elements of the sericulture process and produces a superior product. In addition, demand for innovative textiles can be met by the company's silk technology platform to generate novel fibers tailored to consumer tastes and needs. As a new fiber material, spider silk has numerous potential markets in addition to non-technical textiles, ranging from medical devices to military applications, sporting goods, cosmetics, and composites.

Technology and Product

Silks are protein-based fibers produced by many arthropods, most notably silkworms and spiders. Silkworms produce a single variety of silk while spiders produce up to seven distinct kinds per individual, with the chemical composition and mechanical properties of each tailored to a specific use. Dragline silk is one kind of spider silk that is fabled for its breaking strength, extensibility, and toughness. Due to their lack of availability and impressive mechanical properties, research over the last thirty years has sought to discover a method to produce spider silk in commercially viable quantities.

Whereas silkworms can easily be farmed over large areas and grown in dense environments for the collection of commercial quantities of silk, spiders have completely evaded mass cultivation. Unlike silkworms, spiders are carnivorous, territorial, and produce low quantities of silk per individual. Refactored is using recent advances in genetic engineering and synthetic biology to redesign the natural spider silk gene to enable expression of silk protein in a recombinant host. This allows the production of silk from natural and inexpensive materials without the need for live spiders. The material is spun into fibers with a synthetic spinneret, which is engineered using cutting-edge microfluidic technology. In combination, these methods allow scalable, reproducible, and "green" manufacturing of high-performance spider silk fibers[22].

The company's technology platform allows the creation of precisely designed silks from protein sequences to spooled fiber. This process begins by determining a protein sequence from natural silk sequences in available genomic databases that will result in the best polymer to fit a customer's need. The chosen protein sequence is then converted to an optimized DNA sequence for expression. Starting in 2010, the National Science Foundation (NSF) Small Business Innovation Research (SBIR) program funded the startup's research efforts including the development and testing of several optimization algorithms. The optimized DNA sequence is then synthesized into physical DNA using a proprietary method. The physical silk DNA is transformed into a high-expression yeast strain to produce protein that is then purified. This purified silk is spun using a proprietary biomimetic microfluidic device developed through research funded by the same NSF SBIR program. The resulting fibers are wound onto a spool, tested, and fed into the downstream process to create textile-quality recombinant silk fiber for use in the apparel industry.

The company anticipates three general classes of products - non-technical fibers, technical fibers, and bulk silk protein. The non-technical fibers are additional textile products differentiated by gene sequence, properties, and processing techniques. Technical products are developed by leveraging the unique mechanical properties of spider silk for potential markets that include ballistics applications or next-generation composites for automobiles and aerospace vehicles. Bulk silk protein applications stem from waste silk protein such as hydrolyzed silk powder for use in the cosmetics industry.

The clearest competitor is the existing sericulture industry that supplies a hundred percent of the requirements of the silk textile market. Like the silk industry, the man-made fiber industry composed of nylon, polyester, rayon, and synthetic fibers is highly fragmented with almost half the global production in China. Other competitors are companies attempting to produce silk via silkworm genetic engineering to produce new varieties of silk, and companies using molecular approaches like Refactored. In the first category, no company has a product in the market, however, this technology has the potential to enter the market using the existing sericulture infrastructure and production methods in low-wage countries, although these silkworm genetic engineering methods have a long cycle time to develop and test a new gene design. In the second category, one company using recombinant genes from goats to produce a short silk fragment is successful even if their material never can replicate natural silk's mechanical properties. Another startup is trying to produce a similar material produced in E. coli for medical uses such as drug delivery, films, surface coatings, gels, and fillers.

The IP landscape consists of silk gene DNA sequencing, recombinant silk gene construction methods, recombinant protein expression strains and methods, and fiber spinning techniques. Silk gene DNA sequences have been studied and patented mainly by academic groups. Many such patents remain available for licensing with other sequences soon coming off patent. The sequencing of additional silk genes is becoming cheaper and faster, enabling access to a wide variety of new sequences from the forty thousand species of spiders worldwide. Recombinant DNA construction methods are an active area of research in synthetic biology. Groups who create such silk producing strains commonly file patents to protect against direct copying. The methods used to produce silk material from these organisms are usually protected by trade secrets. Biomimetic silk spinning methodologies are studied by a small group of researchers and some patents exist although none of them affect the more traditional wet spinning methods Refactored uses to access the non-technical textiles market.

Refactored has developed IP around strains to produce recombinant silk protein, and devices for controlling the polymerization of fibers. Venture capital investment has provided funds to aggressively file patents to protect current and future inventions. The firm's IP strategy, encompassing the four technologies mentioned above, focuses on protecting each level of the technology platform for delivering silk fibers. It is a blend of patents for technologies that could be discovered by other researchers, and trade secret protection where proving infringement is difficult.

Business Model and Execution

The total funding received from the NSF SBIR program in the period 2010-16 will amount to approximately $1.4M. This includes funding for joint research with the NSF Engineering Research Center, University of

California at Berkeley. In the formative stage, investment in the form of convertible notes from angel investors was used for working capital. In 2011, the company closed a venture capital investment round of more than $4M to accelerate the development and launch of non-technical textile products. Refactored plans to sell its silk in yarn (5-10 fibers twisted together) or fabric form to designers for use in their textile and apparel products. Initial sales will be to apparel companies who want to replace their existing silk suppliers with an unbranded version of the company's silk at competitive prices. As this revenue stream is established, the company will begin branding Refactored Silk as the superior alternative to both natural and synthetic fibers.

Experts in the textile industry such as the above-mentioned Sue Levin and Greg Haggquist along with the former CTO of Nanotex, a nanotechnology-textiles startup, have provided unique insights to develop market strategy. These individuals have advised that upon providing a sample of textile-quality material, the company will be poised to rapidly enter the market. Refactored's near-term go-to-market strategy focuses on being capital efficient and responsive to feedback from early customers.

Initial runs of silk protein will be produced by outsourced fermentation of their yeast strain by a facility that can support a thousand liters-run and help design and fine-tune the protein production and purification process. Larger contract manufacturers have been identified for pilot-scale production. The outsourced protein is spun into fibers in-house on a continuous spin line to create spools of prototype fibers. The company has selected initial customers to examine and use the prototype fibers and fabrics to provide feedback towards purchase. The ability to rapidly iterate around customer needs will allow it to increase speed-to-market with a textile quality fiber that will then be woven into fabric swatches, the common medium for fabric sales to designers, to secure initial orders. Converters are utilized to turn silk yarn into fabrics and garments, or a customer can choose to pull the product through their preferred supply-chain.

Product development is planned to be accomplished in three phases: in the first phase completed in 2011, proof-of-concept and key aspects of the technology were demonstrated with funding from the NSF SBIR program and angels. The second phase, 2012-2013, funded by NSF SBIR and venture capital investment, involved prototype development. This entailed the creation of a recombinant silk fiber suitable for use by a textile customer, a manufacturing plan based on outsourced production, and establishing relationships with a few initial customers. In the third phase, 2014–2016, it is expected that the company will begin to generate revenue through product sales to initial customers utilizing outsourced protein production capacity and begin construction of a demonstration-scale production facility (approximately 300 tons/year) using additional venture capital investment and project financing.

Refactored plans to generate revenue through sales of silk fiber and fabrics with initial production expected to begin in 2015 utilizing a demonstration-scale facility with sufficient fermentation capacity and spin-lines to convert protein into fibers. Cost estimates will be arrived at based on building similar scale plants used for bioethanol production or retrofit of an existing bioethanol plant. Actual production costs will consist of media, labor, and operating costs. The firm's production organism can grow on minimal media, primarily glycerol or sugar, methanol for induction, ammonium sulfate for nitrogen, phosphate, and salts. Cost savings are likely to result from the ability to re-sell media components as fertilizers, a common industry practice. Both fermentation and spinning at this scale can be highly automated. Additionally, the processes do not involve hazardous materials or require complex or difficult overhead.

NSF Commercialization Assistance and Impact

Refactored Materials was introduced to Innovation Accelerator (IA) in June of 2010 by a member of IA's network. The relationship between IA and the startup began through a series of meetings at the Synthetic Biology Engineering Research Center (SynBERC) at the University of California, Berkeley. SynBERC is one of fifteen Engineering Research Centers (ERC) funded by NSF. Each ERC focuses on a specific science and technology area. The level of funding provided by NSF for an ERC is between $3-$5M/year for ten years.

IA's Entrepreneur-In-Residence was intrigued by the commercial potential of bio-based materials technology; even more importantly it was clear that the founders sought not to simply continue to research the science but were determined to start and build a successful high-technology company. The NSF SBIR process requires an applicant to think beyond research to begin to assess the commercial viability of research outcomes. Company founders were receptive to guidance and eager to incorporate lessons learned through IA's experience working with hundreds of NSF SBIR-funded startups. Utilizing the resources and networks provided by SynBERC and IA, the three founders gathered input that helped shape its strategy and commercialization path.

It was clear to the startup that additional funding would be required to approach the market with a minimum viable product. The company had begun to receive interest from VC firms through publications, word-of-mouth, and venture forums. As the firm began to short-list ideal funding partners, the founders and IA worked together to think through appropriate deal structures, future funding cycles and funding levels to get to exit, Board composition and other key items of corporate strategy. IA was able to provide robust guidance considering several other similar deals IA had previously been involved with or participated in. This feedback provided a benchmark against which the startup could ascertain the benefits and potential pitfalls in negotiating deal terms. IA continues to assist the company with negotiating a term-sheet for a substantial funding round with a partner that has the capacity to satisfy the firm's funding needs through their targeted exit.

The startup's needs as identified by IA converged upon the company's R&D program and its efforts to transition to the commercial market. IA's work with Refactored has focused on transitioning from R&D to commercialization planning to obtain NSF SBIR funding and bridging the gap between grant funding and venture capital funding. Refactored Materials was one of four featured startups at the "IA @ Stanford" event. The startup participated in a case study providing insight into startup challenges as well as NSF opportunities and support. This opportunity allowed Refactored to refine its go-to-market strategy as well as providing Innovation Accelerator insight into technology pain-points that might be addressed by NSF SBIR-funded companies.

IA activities and outcomes are summarized in the following Table.

IA Activity	Outcomes
Startup Overall Engagement	37months; 20+ hours of interaction; 75+ e-mails sent to and on behalf of Refactored Materials
Operational Assistance	Identify funding opportunities; provided financing introductions and guidance
Introduce Potential Customers/Partners	Sherman-Williams; Sigma-Aldrich; Monsanto
Find Investors	Allen and Company; In-Q-Tel
Identify Domain Experts	Kevin Dewalt; David Milligan, Advent IP
Provide Proposal Guidance	Assist with NSF SBIR Phase II commercialization plan
Feature at Trade Shows and Special Events	Featured at the "IA @ Stanford" event

Where was this startup at the end of federal government funding?

In August 2011, the company closed on their $5.1M Series A round from Lead Foundation Capital, and in April 2014 closed on their Series B round of $32.5M again from the same VC firm that participated in the Series A round. NSF's matching grant of $500K enabled additional technical development towards a market ready demo. Series A funding was to support technical development of a scalable process for manufacturing the product. The result of this technical development was the Series B financing to begin manufacturing and sales of the product. Instrumental in this financing were letters of intent from leading customers (large brands with market caps of over $10B) to purchase and use the company's silk in their consumer products. There were no products or processes in the market at the end of this federal funding. The number of employees ballooned from three to thirty-four during federal funding, and revenues grew from $0 to $1.3M. The company is amid scaling production in preparation for product launch and growth. The commercial development to support this has included hiring a Chief Marketing Officer from the target customer base, building a competitive book of potential off-take partners, and developing a vertically integrated strategy for product development and launch.

Where's the startup now?

In May 2014, the company formerly known as Refactored Materials, Inc. changed its name to Bolt Threads Inc. the firm engages in the manufacture of fibers and fabrics. Advances in biology are already revolutionizing a series of industries, from food to medicine to fuel. Now Bolt Threads is set to bring textiles into the fold. Its team is creating new bioengineered materials, starting with engineered spider silk that can deliver any combination of specific attributes. The textiles industry is particularly ripe for an overhaul. The last true materials invention came when the chemical industry created and commercialized polyester in 1953.

The company's first application employs gene sequencing technologies to replicate spider silk production at a commercial scale. Thanks to its complex internal structure, spider silk is one of the strongest, lightest, and most promising materials. Apply those properties to textiles, and you get a fabric with hundred times the strength of reinforced steel but that is as soft and flexible as the most comfortable fabrics. What struck investors about Bolt Threads' founders, is that they were visionary biologists seeking to transform an antiquated industry. The concept is simple: Start with the best materials from nature and use bioengineering to create new fabrics that address specific needs and uses. The execution, however, is massively complex. Many companies have made similar breakthroughs only to falter when moving from the lab to market. Crucially, the engineering team at Bolt Threads has been successful in avoiding the pitfalls that typically sideline promising technologies.

Bolt Threads has focused on large-scale commercialization and scalability enabling it to develop technologies that could withstand the rigorous path to market entry. Instead of a dedicated manufacturing plant, which would have required hundreds of millions of dollars, this startup has elected to pursue partnerships to ensure that it has the flexible manufacturing capacity to meet consumer demand. The company used gene sequencing to identify thousands of different proteins that could produce specific attributes in spider silk. This intellectual property represents a competitive advantage that could be applied across a range of industries, from medicine to airplane production. Rather than competing solely on cost, Bolt Threads offers high-end quality products for the high-performance segment. Its technology also has important environmental benefits. Its focus on natural materials alleviates the need for petroleum, which is used to manufacture many synthetic textiles. By producing silk in the lab, the company does not have to rely on thousands of silkworms, a species struggling due to the detrimental effects of climate change. In the next several years, Bolt Threads will ramp up production to manufacture tens of millions of pounds of fabric. More important, this breakthrough represents just the first in a wave of technology-driven products from natural materials.

In 2017, the company sealed a partnership with Stella McCartney, a world-renowned brand known for its dedication to eco-innovation. They plan on unveiling their first collaborative product, a one-of-a-kind custom dress made entirely of Bolt Microsilk™ at the New York Museum of Modern Art's upcoming exhibit "Items: Is Fashion Modern?" this year. Stella understands how important the brand's material and resource choices are, and how those impact the environment. The company knew it must work with this brand when it heard Stella was willing to pioneer new sustainability techniques in fashion:

"We're excited by risk. We're excited by thinking outside the box. We think that's modern. And I think the fashion industry is supposed to be modern, and I find it so extraordinarily old-fashioned at times that I can't really get my head around it. The fact that we're even having this conversation, it seems medieval to me. This is something that I've been on a personal journey to find, for much of my career, and I just feel like there is finally a new opportunity to bring so many industries together and for them to all work as one for a better planet. It is a truly modern and mindful approach to fashion."

The museum piece, a custom knitted one-of-a-kind, gold dress, is a modern interpretation of the classic shift dress and is made entirely of Bolt Thread's signature protein-based yarn inspired by spider silk. The dress will be made using this brilliant golden Bolt yarn. This partnership will pave the way for a future of environmentally friendly luxury fabric innovation. Throughout 2017 and beyond, the two partners will continue to announce new initiatives that will change the future of apparel production. The company is inspired everyday by the amazing materials they work with. It is driven by the desire to turn these materials into unique products. Bolt Threads is led by world-class scientific and engineering talent, as well as experienced executives from the technology and apparel industries. In the not-so-distant future, we all could wear clothes or sleep in tents made of synthetic spider silk.

The company uses combinations of proteins to create new types of materials. Building the material from proteins allows it to adjust the properties of the material, such as for strength, stretchiness, softness, or comfort. Recently, Bolt Threads again made headlines when it announced a $50M investment round and a deal with Patagonia to develop products made with Bolt Threads yarn.

Case Study Questions

1. Commercializing basic research funded by NSF at an Engineering Research Center - assess market opportunity/potential and value proposition
2. Identify risk elements and how to manage/mitigate them?
3. Additional funding/growth strategy
4. Barriers to entry
5. Right business model?
6. Identify areas where startup needs help
7. IA approach the right one?
8. Startup valuation exercise
9. Take-away(s)

17   Plastics Transmit Radio Waves

Technology Sector: Wireless
Startup: netBlazr Incorporated
Website: www.netblazr.com
Location: Boston, Massachusetts

Federal Funding Timeframe: January 2013 – June 2017
Funding Amount: $0.88M

This narrative is a story of: Sales/bootstrap growth; serial entrepreneur: company focused on wireless internet access – this grant for transparent mesh antennae placed on windows, not their core business; turns out not to be a viable product

Startup Formation

Brough Turner and Jim Hanley are the founders of netBlazr, a Boston-based company they started in 2010. Brough, an electrical engineer, and a graduate of MIT, co-founded Natural MicroSystems (NMS) in the mid-80s. NMS went through several business models in the late 1980s until it hit upon a computer telephony play that was successful and went public in 1994. Brough later started two other ventures. One produced a successful product but never became a standalone business, and was eventually sold for $2.9M, after an investment of $4M! The other ended up a penny stock company. In 2009 with his extensive wireless business experience and frustrated by the broadband duopoly that exists in the U.S., he decided he wanted to make a difference. In his travels in Sweden, Japan, and Korea he had found much faster Internet access than in America, and much cheaper.

"I'm not a politician and I don't have the resources of the Google guys, so the only way for me was to come up with a disruptive business and make an end-run around the large companies and the government. But it had to be highly profitable – without large profits I would make no impact."

Brough is a technologist, has dabbled and succeeded in many different endeavors, is good at strategy, product marketing and technical sales support. He can do operations but prefers new technology development - "I find partners to do things I am not good at or don't enjoy." Jim met Brough in 2010 through a mutual friend. Jim has decades of experience in technical marketing, sales, and general management in the wireless industry, and earlier in his career found work at a military contractor too staid. In 1993, a mobile data pioneer, he worked at Microsoft in mobile strategy development while obtaining his MBA from Michigan. Jim was the fourth businessperson to join Nextel in the northeast as a product manager when it was a startup. He became a general manager (GM) after five years – by then the company had become too big so he left to join a venture capital-funded turnaround. Jim loves startups, grungy and small, with great work environments, and in the decade 2000-2010 worked with multiple startups as a quasi-consultant and co-founder, helping to turn a few of these around. His turnaround efforts also included working as division GM for what is now one of the biggest fiber networks in the northeast.

Jim was involved with a startup when he first met Brough. They talked back and forth for a few months, got to better know each other, and developed a strategy to execute on Brough's vision for a product and service incorporating radical changes, enough to disrupt the wireless broadband duopoly of Comcast and Verizon. Originally, there was a third founder with a Harvard MBA, but with only netBlazr sweat equity and uncertain prospects for a steady salary to support a young, growing family, he was forced to find work elsewhere after a partial stock buy-back agreement that was amicably negotiated. He remains on the Board. To get started, Brough invested $50K and Jim $20K in convertible notes. A few months later, after preliminary wireless network testing and validation they raised an additional $275K in convertible stock from family members and friends, one a well-known entrepreneur with multiple successful previous exits.

Technology

In seeking a path to disrupt the broadband market, Brough spent 4-5 months investigating access to fiber right-of-way – this turned out to be a dead-end, so the next option was a wireless end-run. An Eastern European company, MikroTik, and a Silicon Valley company, Ubiquiti, were selling low-cost radios in the developing world and in rural areas of developed countries. In rural America, there are now thousands of wireless internet service providers (WISPs) using this low-cost unlicensed spectrum technology.

"Can we find a way to use these radios in urban areas where there's lots of interference, for example, with short-range directional links and multiple repeaters? And can we structure a business or service where people want our technology on their roofs or in their windows? We arrived at a novel way of using existing technology and untried, but clever, business models that avoided the need for expensive roof-rental or cellular-tower rights."

Brough and Jim built a pilot network in 2011-2012. They connected two-dozen Boston-based companies and developed software to deploy the network. The technology worked, the business model brought in ever more rooftops, they established that they could compete, but one obstacle remained - the need for professional installers[23]. The initial business is viable for tens of thousands of customers but not for hundreds of thousands or millions of them. Short directional links must be re-aimed as customers come and go, so installers will increasingly be called upon to re-aim and re-position existing radios. Reconfiguring the network this way costs time and money. The business had no need for significant technology development or licensing – it was about using existing radio products in novel ways. netBlazr researched a possible solution to eliminating the re-aim and re-positioning requirement, to auto-aim without professional installation. They filed two patent applications, but the need for further research was evident. Who would pay for this R&D effort? They stumbled upon the National Science Foundation (NSF) Small Business Innovation Research (SBIR) program.

"In 2011, we entered the MassChallenge Business Plan competition. As a semifinalist, we were given free rent for six months in a downtown incubator facility. In the next cubicle, there was this startup being funded by the NSF SBIR program – they told us about SBIR; another biotech startup in this same facility had SBIR funding from the National Institutes of Health. I had heard vaguely about SBIR but didn't know enough, I have never asked the government for funding. It was all new to me."

The team decided that SBIR grants, if available, could fund this R&D effort. Turns out that the design of a transparent antenna is critical, but Brough was not an antenna designer. They searched the country to find suitable research partners. They found a professor at University of California San Diego, who referred them to potential collaborators at Drexel University and then at the University of New Hampshire (UNH). The mathematics and simulations used in Phase I of the SBIR project proved a solution is possible. This funding covered basic calculations, electromagnetic simulations, and helped identify transparent materials with adequate conductivity. The product developed under the NSF SBIR project would eventually require Federal Communications Commission (FCC) approval and would be best deployed on windows. The two founders have considerable experience with FCC rules for unlicensed spectrum usage and plan to engage outside labs for product testing. Two years from now, the goal is to have developed a prototype transparent patch antenna wireless mesh network node that a consumer would be willing to hang in their window that automatically aims and re-aims on a packet-by-packet basis. This prototype will serve as the basis for raising venture capital (VC) funds in the $4-5M range. A successful SBIR-developed product will make the existing netBlazr service business truly disruptive and will foster a new products business selling mesh network nodes to community networks and service providers around the world. Although IP generated will lead to further patents Brough says –

"I've been involved in patents for thirty years. The reality is that patents are less of a barrier in this business, but we still seek them because some investors favor startups with patents."

Customer Discovery and Business Model

Other companies provide wireless internet services in urban areas. Towerstream, for example, provides service in twelve major cities focused on business customers using point-to-point links from tall buildings. It is an expensive model. netBlazr's model involves a dense mesh of interconnected radios with short hops between low-cost routers and multiple paths for redundancy, so costs are dramatically lower. Cost of providing service ranges from $200-$500 for installation. The initial business model focused on business customers with radios operating at frequencies of 5 gigahertz (GHz) and limited 24GHz mounted in window frames, but this led to wiring and workspace issues.

"We find roof-space we can use for free to place three or four 300mm antennas, typically on buildings such as apartment/condominium complexes that don't have cellular gear already on their roofs. We explain to apartment managers the extra benefit for tenants who get higher speeds for much less cost. The objective initially is not to make major profits but to achieve cash-flow breakeven and have enough of a profit margin to invest in further growing the network. In the beginning the focus was on businesses because they are most beholden to the duopoly. We thought every business would be a relay point, but today we've found building-rooftops are the best relay points whether commercial or residential. Building managers like to have an alternative to the duopoly."

Early adopters were companies in Boston – people who trust this startup and the team behind it, and well understand the benefits of their service offering. Many pay exorbitant fees for bandwidth and are thus willing to try out netBlazr as a second and/or backup service. Some find they pay much less using this company as the primary provider with one of the duopolies serving as backup. The initial business model's focus on business-to-business (B2B) and not residential customers turned out to be wrong – selling to businesses is very expensive. So, they added residential services, even though they thought cable companies had 'residential' covered. Soon they were adding residential subscribers at a fast clip and obtaining rave reviews on Yelp and other consumer web sites. Interestingly, rave reviews on Yelp also brought new business customers, without the need for B2B sales. The total subscriber count is now approaching 300, split between residential and business customers. There are more residential customers, but revenue is skewed to business, with mostly business usage during the day and residential usage at night. The company foresees little near-term competition in Boston. Monkeybrains in San Francisco has over 2,000 subscribers, and a similar business exists in Washington, D.C. As far as the duopoly, Verizon did not react to Comcast taking their broadband business away until they had lost 20% market share, and now Comcast is the dominant broadband service provider. It will be some time before netBlazr is noticeable to either. Problems to resolve are mostly technical in nature - issues of interference, network management, and provisioning new customers. Regarding network speeds on offer – the duopoly advertises high speeds but in actual practice netBlazr can beat these duopoly speeds, with many different service offerings both for up- and down-links. Urban areas with their much higher density require the right view - tall office buildings work well as do condominium or apartment buildings, but any building a few floors higher than its surroundings is a good relay point. Over the next twelve months the challenge is to test out and fine-tune the business model, properly manage scale-up and expansion in Boston, and then identify ideal large metro cities in other states to rollout service before competitors arrive.

Financing

netBlazr started in 2010 with $70K in founder funds and sweat equity. The two founders pitched belief and expertise in their technology and their track record and business model to family members and investor friends for an additional $275K to implement their pilot network in Boston. They estimate they will need an additional $300K from current investors to accelerate the build-out of the wireless network in Boston - much less capital is required for network reconfiguration and installation. A year from now they will be well-positioned for growth capital of $4-5M from VC or private equity firms, to use for network rollout in five more cities. Efficient execution of a properly fine-tuned business model for these markets will result in a highly profitable business, with a price umbrella provided by the duopoly. In conversations about funding sources the SBIR program had come up briefly.

"We heard it was a big hassle, lots of work, low chance of success and besides, we were already on the VC treadmill and were focused on that. After a year of this VC distraction, we decided to give SBIR another look, decided to give it a shot."

Having failed at their first attempt with the NSF SBIR program, they sought advice from a broader set of knowledgeable people and experts at universities familiar with this program. They succeeded in their second attempt. Professor Nick Kirsch from UNH served as a consultant to help prepare the NSF SBIR grant proposals. Phase I funding brought in $150K. netBlazr is well-aware that sans the automatic-aiming feature the cost to reconfigure the network will be labor-intensive and therefore expensive – it is why the SBIR-developed product is important. It would vastly increase the addressable market; from being a niche-provider the company would be able to target greater than fifty percent rather than perhaps ten percent of the market.

Market timing considerations led them to not delay the Phase II grant proposal. They submitted it at the first opportunity which came six months after their Phase I started. In 2014, they were awarded Phase II funding for $720K. The founders will consider raising additional private capital only when they are confident enough to deploy their current business model in another city, or when more Phase II milestones are accomplished and when capital is then required to build prototypes of the NSF SBIR-developed product. Moving forward, Jim and Brough envision two funding paths – the first to execute a new business model around the product developed via the SBIR-funded research, and the second to scale the current service-provider business. Regarding company valuation, the founders had this to say –

"So far it has all been convertible notes with friends and family. VC negotiations are always tough, but we've done this before. They must like and trust our team. We present burn rates, make reasonable projections, this is more art than science, we let the VCs consider our team, the opportunities, the business model, the revenues – this part of the business is worth this much, this other part that much, we will have a product prototype, we name a price – say pre-money now $2-3M, and maybe willing to negotiate and adjust it a bit."

Company Growth

The Board of Directors consists of Brough, Jim, and the third founder, Jason Orgill. Jim, the CEO, and Brough, the CTO, make the decisions. Colin Zwiebel, in-charge of network deployment, and George Kontopidis, along with the two founders make up the core team. Brough hired George in 1989 in his earlier company. George has held various high-level positions over the last twenty-five years – he is a technology management expert who has recruited large numbers of engineers, and in the past grown many engineering departments, one a 250-person team in the mid-1990s. George works twenty hours a week on the NSF SBIR project. Two years ago, George recruited Colin from Olin College to primarily handle day-to-day operations of the wireless network service. Now, a new person has been hired to take Colin's position, so he can focus on building the embedded radio on the SBIR project.

How do you retain key people?

Brough - "With stock options, decreased but reasonable pay, giving them exciting things to do, work on something cool - disruptive anti-establishment David v. Goliath scenario, smart people to interact with, give young guys lots of responsibilities. Colin is two years out of college and was running a network!"

Jim - "The amount of bootstrapping required, working for no salary for more time than we expected, but we must keep it going, that's our nature, most would have blinked long ago. We are still passionate, we persevere, do whatever it takes, the classic entrepreneur, get your hands dirty. Many people work at startups well-funded by VCs and at regular salaries – we started in a garage with rats (Brough corrects him - not rats they were just mice), we run up and around rooftops. We must stay focused, eventually it will come to us; in everything do whatever you have to, recruit people with complementary skills, different viewpoints, we end up doing things comfortable, some uncomfortable, but we must do it."

Brough – "In my first company I had three partners with different skills; I don't look for loners but try and make up for my deficiencies; need diverse and different points of view; bad if you can't stand each other; judging people – there are two parts to interviewing and recruiting, deciding who to hire; and setting the tone for company culture. I'm not an expert at interviewing people. George, he'll ask questions that embarrass me; we need complementary personalities."

Jim – "Share the passion; it's not what the person's resume says, new people need to understand what they are getting into. Open kimono – let people test-drive us, hangout with us, work them for a few months; drink Kool-Aid with us; if there is something magical between us, it will show up; we are always recruiting, looking for technology talent, telling our story; old-fashioned way sometimes works – networking, referrals; if people don't work out – sometimes the problem is us, sometimes not; forget exhaustive interviews, try trial internships; getting rid of people – not easy to fire, if not working out they know it, can't hide that; sales people that don't work out usually pushed out early."

Brough – "Lots of interns in MassChallenge space; many interesting events to attend; interning good opportunity to study them; Boston's a rich place for people skills. Ans there are plenty of research universities in the area within driving distance."

Do you plan to manufacture products in America?

"Not in the next two years. SBIR product quantities will initially be hundreds and then thousands a month and can be done in America. Even for larger volumes Brough and I have enough connections with U.S. contract manufacturers, but for consumer products because of the high volumes involved, you'd have to go overseas, to the Far-East, to engage contract manufacturers in those regions."

Key Relationships

The Wireless Internet Services Provider Association (WISPA) has about 850 members of which 5% are non-U.S. firms. The founders use WISPA to track industry issues – they have two trade shows per year. One founder attends this trade show every other year to examine vendor products, to scope the landscape, and meet people. Face-to-face meetings are important although most interaction is done via email and telephone. The online information WISPA provides is invaluable. MassChallenge provides many connections in Boston, people you can call if you have investor-related questions and questions on SBIR. It helped Brough get reconnected after many years of work that had focused out-of-state and overseas. In terms of business development, the Boston Mayor's Innovation Office and property management groups have been useful.

"Of course, NSF is important – it is strategic, gives us the ability to do new product R&D that would otherwise languish in a garage. Now there's research to do, more structure, direction, and funding; it's the difference between doing it and a promising idea sitting on the shelf. NSF gave us great credibility, provided cache, good housekeeping seal; the Phase I workshop media session was useful. We made several good contacts in the radio/wireless business and with antenna-related people at the NSF workshops."

Are university facilities important?

"Yes, for antenna expertise we looked at MIT but ended up at Lincoln Labs and the Worcester Polytechnic antenna group. Professor Kirsch at UNH has been invaluable and UNH has an anechoic chamber we need for antenna testing. The New England Software-Defined Radio Consortia monthly newsletter is useful. Ideally, partners in and around Boston is what we prefer; we can visit easily. Later, for the SBIR-developed transparent antenna wireless mesh node, we'll need a series of partners for manufacturing, beta testing sites, and distributors to list our products. Packet-by-packet aiming will be a fundamental game-changer for building wireless networks in urban areas; we'll then become a product company."

The Startup Experience

Brough and Jim, from all your multiple entrepreneurial experiences, what advice would you give a budding entrepreneur?

"If someone is interested in doing a startup rather than a 9-5 job, we'd strongly encourage it, especially if they're young, unmarried, with not too many responsibilities, just go for it. First figure out pieces of the business model before you scale and get going. VC money can tempt you to spend millions quickly before you've figured out the business model; you spend money on something that may not work. In the service-provider business, get customers ASAP, understand their needs; real live customers – they may yell at you or love you; you can quickly and cheaply get a business going; products, a different problem – you need to understand the ecosystem, competition, you must get in and do it, it takes time and likely more money than you have."

"Solo startups are hard - even two people often isn't enough; need minimum three who share your passion; someone to hold the torch all the time. When starting out, you need not have all three founders 100% committed. VC model is fine when you are relatively confident that you can scale, and you know how. This model is often premature where the business model is not yet proven; simply throwing money at it doesn't work. One should try and do it without wasting too much money. VCs get way more credit and attention than they deserve, a fit maybe for 10% but they think they drive 80% of the innovation in the U.S."

Does luck have anything to do with it?

"Luck comes to those who have thought of a million options, then they'll recognize when opportunity comes; luck is nice but if you've already thought of all viable options, then how much is luck, how much is preparation? Luck applies more to the traditional VC market, a roulette wheel, looking for one home run; in the mobile apps' world and dotcom exits - luck plays more in that world."

What about your personal qualities?

"A certain high-level of confidence, believing in yourself, having a vision, and believing you can realize it; some level of persistence, need to be flexible; perseverance, ugly side stubbornness; work ethic – get more done, be relentless, keep pushing even if many doors are slammed in your face. If you can't take those ego slams because most of the time you are wrong, then you are in the wrong business. Lots of self-motivation, keeping at it; confidence is infectious, lead the charge. Know when to quit, know when to make a change and pivot. If nothing is working, you have core business model issues, or the economics of the business is just too difficult, better to quit than raise money."

Where was this startup at the end of federal government funding?

The original technical objectives were not met. The underlying concept is still viable and the research that was accomplished sets a good foundation for an alternative product or products, but the original proposed product has not yet been shown to be economically viable. Three US patents were granted: US 8,667,148 - Minimal effort network subscriber registration; US 8,892,048 - Transparent antenna; US 9,537,216 - Transparent multi-element antenna; there have been no revenues thus far. At the start of this federal funding there were six employees and at the end of this period there was eighteen.

Since beginning work on this federally funded project, there has been a dramatic increase of interest in fixed wireless broadband technologies and capabilities. A new entrant called Starry Internet, headquartered in Boston, announced significant investment in its business that will leverage new radio spectrum to deliver high-speed internet service directly to consumers. The acquisition of Webpass Inc., a direct competitor of netBlazr, by Google Fiber was further indication that the broadband industry is looking for ways to leverage wireless technologies to deliver "last-mile" broadband services.

The team strongly believes that, in the coming years, the transparent antenna node technology which netBlazr is developing, will be a very attractive complement to the capabilities that the industry seeks. Recognizing that productization of the transparent wireless network node will require a substantial additional investment in radio technology, the company is targeting radio manufacturers. There are other potential partners who could assist in commercialization of transparent wireless mesh node in other segments of the industry including antenna providers and a few innovative service providers.

Where's the startup now?

netBlazr is a locally owned business that offers "nothing but net", an Internet connection that is reliable and affordable and nothing else. No gimmicks, games, or tricks. Just broadband. This company was founded by a group of technology entrepreneurs who were frustrated by all the games the big Internet companies play with their customers. Promotional offers that lead to huge price jumps after they expire, requirements to buy into services not needed, it's all just a way for Comcast and Verizon to squeeze more revenue out of their users.

Case Study Questions

1. Handling a business developing products that require meeting FCC regulations
2. The broadband duopoly in America – can they really be disrupted?
3. Comment on the two-track approach – viable business alongside research funding through NSF SBIR
4. Assess commercial potential and value proposition
5. The scaling factor – what would it take to successfully execute

6. Develop a technology strategy/roadmap for the potentially disruptive part of the business
7. Identify risk elements (technical, team, market, finance) and how to manage/mitigate them?
8. Growth strategy
9. Start-up valuation exercise
10. Take-away(s)

18   Displays You Can Bend and Twist

Technology Sector: Optoelectronics
Startup: Orthogonal, Inc.
Website: www.orthogonalinc.com
Location: Ithaca, New York

Federal Funding Timeframe: January 2010 – September 2016
Funding Amount: $1.18M

This narrative is a story of: Lab-to-manufacture innovation model; private equity funding; contracts; strategic partners; IP play; scale-up/production challenges; assistance from the State of New York

The Lab-to-Manufacture Innovation Model

In executing the Lab-to-Manufacture model to deploy our country's manufacturing heritage to rebuild global competitiveness, the Innovation Accelerator (IA) initiated a long-term relationship with the Eastman Business Park (EBP). This partnership would allow startups funded by the NSF SBIR program to fully leverage Kodak's manufacturing facilities in Rochester, New York. EBP has been Kodak's primary film manufacturing hub for over a century. Kodak has now developed this facility into what is today one of the largest, most diverse industrial and technology parks in the U.S. - the only industrial park of its size that was built in a vertically-integrated fashion to support Kodak's R&D and commercialization components. Products manufactured on-site incorporated a broad range of technological advancements over the last hundred years, spanning photography, motion pictures, healthcare, printing, national defense, and document imaging. Today, it is open to the next generation of entrepreneurs, innovators, and employers, making its infrastructure, captured in the figure below, available to help accelerate middle-stage technology companies, and allowing them to develop their businesses from the lab-scale prototype stages of innovation to the later stages of production and commercialization. In this figure NYSERDA is the New York State Energy Research and Development Authority and NY BEST is the New York Battery and Energy Storage Technology Consortium.

The suite of test, validation, prototyping and pilot manufacturing capabilities available at EBP are specifically suited to help accelerate commercial deployment of technology in the functional films, energy storage, and biomaterials segments - all critical elements of the nation's future advanced manufacturing economy. EBP can provide resources to tackle our country's manufacturing challenge with an experienced workforce to train the next generation of manufacturers with technology skills for the development and manufacturing of future technology products. With over a century of deep technical expertise and infrastructure in the U.S., the Eastman Business Park can be viewed as a national model in providing existing assets to support innovative technologies by removing risk and increasing the predictability of product maturity from concept to full-scale production. IA is working to make more broadly available the unique research, innovation, and skilled workforce capabilities that exist at EBP, and create an infrastructure for manufacturing innovation to ensure that the next generation of processes and products not only will be invented in the U.S. but scaled up and manufactured in the U.S. as well. This is both critical to the nation's economic future as well as to solving some of the world's biggest sustainable development challenges at this crucial juncture.

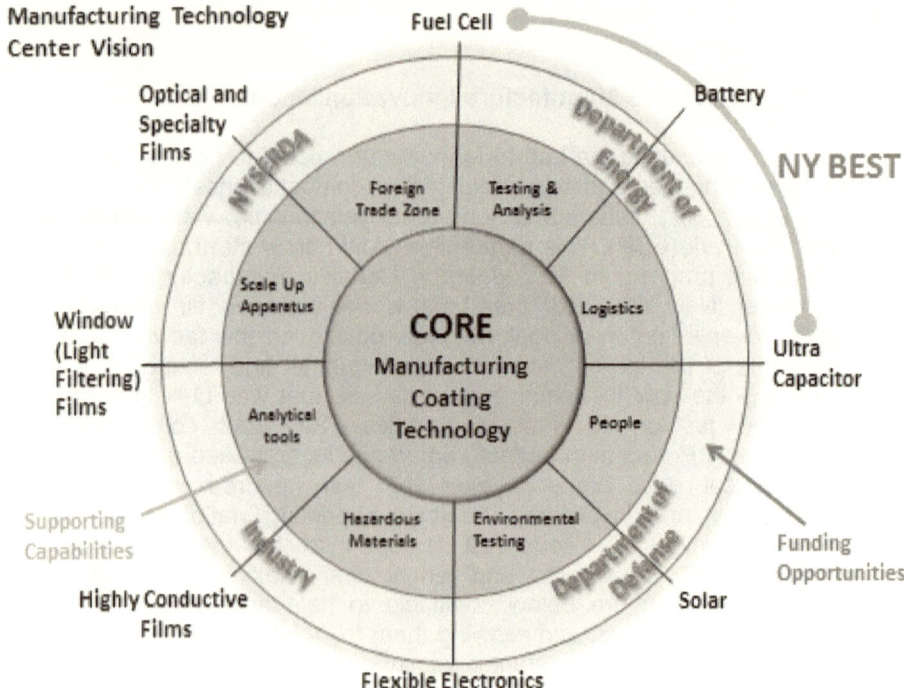

To compete globally, the U.S. must nourish IP beyond the lab from prototyping and proof-of-concept to commercial products, especially with a majority of 'new economy' opportunities in energy, clean-tech, consumer electronics, computing, functional printing, and biotech requiring insights in materials science and chemistry, pilot testing, and infrastructure support. It is about preserving American manufacturing strength globally, about preserving a manufacturing asset and skilled workforce that would take billions to replicate, about refusing to cede whole new industries to other countries. It is about jobs and finding a new path forward in the evolving 'innovation economy'. We need to monetize American innovation by utilizing existing assets, leveraging the existing workforce and intellectual capital, preserving capital, reducing risk, and improving predictability. The Lab-to-Manufacture model and the IA/EBP partnership will facilitate this, helping our nation's best and brightest technology startups become profitable, self-sustaining, job-creating enterprises quicker.

Orthogonal along with four other startups funded by NSF/SBIR now have either their manufacturing facility or are fully located at the Eastman Business Park. This allows them multiple deposition capabilities, bench-to-manufacturing scales, and ready access to a knowledgeable, experienced manufacturing workforce. There are obviously many technology incubators, accelerators, and economic development entities but they are mostly upstream research/business facilities with little or no access to testing/development or pilot manufacturing capabilities. The closest comparison would be a state-funded manufacturing site, focused on photovoltaics for example, in China, or some of the large German facilities such as the Bayer campus. EBP is a 1,200-acre technology and industrial campus that includes seventeen miles of railroad track, 16M square feet of manufacturing, lab, warehouse, and office space, 50M gallons per day of industrial water supply, and self-generated utilities with a 125MW power plant producing electricity, steam, and chilled water.

Company and Team

Orthogonal was founded in 2009 to commercialize research done by professors George Malliaras and Chris Ober at Cornell University. The technology involves a fluorinated photoresist platform that enables the direct patterning of organic electronic and other chemically sensitive materials using standard photolithographic equipment to advance the field of electronic display manufacturing. Its proprietary photoresist solution allows the direct patterning of a wide range of organic electronic materials for organic light emitting diodes (OLEDs) and flexible display applications. The initial focus is to sell the photolithography chemicals to display manufacturers.

The company acquired an exclusive license from Cornell University to scale and sell organic electronic compatible photoresists, and to develop processes to be used by electronics manufacturers to produce high-quality consumer products. The company has two core competencies - photoresist development and organic electronic manufacturing process development - both crucial to commercialize the photoresist. Orthogonal has laboratory facilities at the Eastman Business Park in Rochester, New York, where it can access the best talent from the display and chemicals industry to develop its products and leverage the installed infrastructure to allow the manufacturing of advanced flexible and OLED displays quickly and without major capital investment. In collaboration with a strategic partner, Orthogonal will soon be mass-producing this powerful, disruptive technology.

Fox Holt, the CEO, has investment banking experience at Morgan Stanley where he managed a $75M private placement for a startup airline after obtaining an MBA from Cornell. He has also advised, started, and been employed by nine different startups. At Novomer, Fox was responsible for the company's first product, an electronics material, and that involved developing a process to produce polypropylene carbonate on an industrial scale.

John DeFranco, Founder and CTO, graduated with a Ph.D. in Applied Physics from Cornell University. His graduate work focused on patterning organic electronic materials and novel device architectures. While at Cornell, John also obtained a minor in entrepreneurship.

Charles Wright is a synthetic chemist who joined Orthogonal after having worked at Eastman Kodak for twelve years where he developed an extensive background in specialty organic and polymer process development, commercialization, and project management. He holds a Ph.D. in Organic Chemistry from the University of Wisconsin.

Sung Hoon Ahn, Product Manager, has nine years of experience working in Samsung's electronic materials group developing materials for advanced batteries, solar cells and displays.

Mark Thirsk, a business consultant, has over twenty years of experience in the chemical industry, working with a variety of materials and processes utilized in wafer fabrication. He also has knowledge of business forecasting, strategic planning, M&A, and technical marketing.

The Board of Directors consists of Fox, John, Andre Edelbrock, Founder and CEO of Ethoca, Cameron Piron, President, Chinook Holdings, and an expert in breast magnetic resonance, and Brad Schmidt, the CEO of Tornado Spectral Systems, located in Toronto, Canada.

Professors George and Chris serve as scientific advisors.

Market Opportunity

Orthogonal makes and sells process chemicals to the display industry to shape electronic circuits beneath the display as well as the pixels themselves. The unique advantage is that its material (photoresist) can be used in organic electronics manufacturing to make displays and circuits that are thinner, lighter, cheaper, and more flexible. There are innumerable ways to pattern organic electronic materials but only Orthogonal has a method that uses the vast infrastructure already in place to make liquid crystal displays (LCDs). The company can potentially sell photoresist to many customers for a variety of applications. The target product is a flexible, OLED e-Reader for education using organic thin-film transistors that would be patterned with photoresist from Orthogonal. There is an increasing need for innovative technology in the classroom, and especially for a revolution in how textbooks are distributed and consumed. In the U.S., modern technology and distribution models are starting to change the entrenched textbook industry. For children, the benefits of carrying around a single display instead of three or four heavy books are quite evident. While existing LCDs such as iPads are colorful and responsive, the e-readers themselves are expensive, heavy, lack ruggedness, are prone to breaking, and are power hungry, all of which makes for a slower widespread market adoption.

e-Reader displays that can be rolled and are lightweight made on plastic films using organic electronics are being developed but have not yet been incorporated into commercial products due in large part to the difficulty of patterning organic materials using conventional techniques. Almost all conventional semiconductors used in processors, LCDs, radio-frequency identification (RFID) tags, etc., involve the photolithography process that uses light and extremely toxic photoresist chemicals to transfer patterns onto a surface. The traditional photoresists used for conventional electronics cannot be used with organic electronics. These toxic chemicals will destroy the semiconductor during the photolithographic process. Organic electronics are currently manufactured in very small quantities using non-standard techniques. These crude methods require huge capital outlays for new equipment and in most applications still cannot manufacture products cost-effectively. Finding less expensive processes is a critical challenge to solve, to leverage the benefits of organic electronics.

Organic electronics are useful in making e-Reader backplanes but also constitute a platform for a broad and growing range of electronic products. For example, OLED displays are expected to penetrate the television market soon. The most exciting applications are organic photovoltaic cells, flexible displays, electroluminescent and OLED-based lighting, organic RFID, organic memory devices, flexible batteries, and organic sensors. The company is focusing on organic thin-film transistors (OTFTs) because of their near-term commercial potential and large market impact in the education market. Organic transistors are versatile, thinner, and more efficient compared to traditional transistors.

A large display manufacturer purchased photoresists from Orthogonal and has successfully tested this product in a pilot-line. Recently, a flexible e-Reader demo made with a photoresist developed by the company was shown at a trade show. A materials company signed a joint development agreement (JDA) with Orthogonal for commercialization of its first organic product. Once the photoresist is ready for mass production, this company will be responsible for the formulation, delivery, and quality assurance of the new product. Other display companies that have conducted initial testing of this material have validated that they are both benign to organic electronic materials and compatible with their photolithography equipment. Another vendor has used this material to pattern conducting polymers and developed a process to use with their flagship organic semiconductor material. The manufacturing process of this new class of electronics must be compatible with the current infrastructure for these devices to proliferate. Orthogonal's technology solves this compatibility issue. Due to the reduced heat requirements and the flexibility of organic electronics, manufacturers can make the next generation of electronics paper-thin and virtually unbreakable, allowing them to last longer, and command higher prices.

Technology and Product

Organic electronic materials are carbon-based substances that exhibit semiconductor properties. These materials can have distinctive properties such as inherent flexibility and solution processing advantages, making them ideal for manufacturing the next generation of flexible displays. Properties can be manipulated by changing the chemical structure of the polymers or the small molecules that make up the organic material, including the ability to conduct electricity, emit light, or act as a transistor. Standard photoresist chemistries are based on organic solvents and aqueous developers. These solvents are generally incompatible with OLED and OTFT materials, making standard lithographic processes unsuitable for manufacturing OLED displays and flexible electronic backplanes. Therefore, alternative approaches are used to pattern organic semiconductors, none of which deliver the combination of resolution, throughput, and yield that is offered by industry standard photolithography.

Orthogonal's photoresist product, by contrast, is based on fluorine chemistry and fluorinated solvents that behaves and performs the same as standard photoresist yet is fully compatible with organic electronic materials[24]. Organic electronics has many benefits over traditional electronics. OLEDs can provide thinner, brighter, crisper displays for electronic devices and can use less than one-fourth the power of LCDs used today. They can use a wider range of substrates including plastics. This allows the electronics to be flexible, thin, and lightweight. Organic transistor backplanes are the most flexible that can be made - this will allow a truly paper-like experience along with the integration of displays into our daily lives where before they would have been seen as intrusive. For over a decade, the electronics industry has been researching the problem of large-scale manufacturing of organic electronics. Electronics manufacturers have wanted to manufacture higher profit-margin organic electronics for several years. The main problem restraining the advancement of organic electronics is their manufacturability and the billions of dollars required to build new infrastructure. Competitive pressures of the global economy make it difficult to justify large capital upgrades to U.S. semiconductor manufacturing plants. In 2005, George Malliaras and Chris Ober started a collaborative research effort to develop a nontoxic photoresist to use with existing infrastructure to manufacture organic electronic devices. Organic electronic production can then leverage both current knowledge of photolithography and the capital spent on existing production facilities.

Initial research conducted by the company successfully demonstrated flexible circuits made using industry standard tools. They showed that such a process is feasible and that organic semiconductors can be fully integrated with a variety of inorganic materials. Current R&D efforts seek to commercialize this photoresist technology for flexible, organic transistor-backed e-Readers with high yields. It focuses on increasing the performance of their current photoresist product using industry standard patterning processes (dry etching) and simultaneously making the material sensitive to longer wavelength light using cheaper light-sensitive materials. In addition, numerous other adjustments are being made to improve the resolution and uniformity. Achieving high yields in the display industry is crucial to a successful product, since costs are extremely sensitive to the volume of defects found in the patterns. Orthogonal is currently producing a suite of nontoxic photoresists in its labs in Ithaca, New York, and selling this to the semiconductor industry. Customers and partners have patterned this resist using a wide range of standard equipment. Compatible photolithography processes extend to vacuum- or solution-deposited small molecules and polymers. This gives customers manufacturing flexibility for current and newly developed organic materials enabling them to find the best combination of organic and inorganic materials for a given device while maintaining a common fabrication platform. A change in functional materials would not cause a change in production methods. Orthogonal resists can also pattern conventional inorganic materials such as silicon, indium tin oxide, and silver. Several applications require patterning metals on top of organic electronic materials primarily for conducting pathways and electrodes. A primary benefit is the high quality of the films after patterning. Spin-coated or slot die-coated films are typically of very high quality compared to printed films of the same material. This is because the materials can be formulated, and the solvents chosen, with performance in mind. With the Orthogonal process, a wide range of organic thin films can be patterned via subtraction, leaving a perfectly uniform, high-performance film, without requiring any reformulation.

Photolithography is capable of high-resolution patterning on a variety of substrate materials. The company has demonstrated this, using many different substrate materials, with high resolutions that are more than is required for most organic electronic devices showing that critical dimensions are well under control and are not and will not be the source of yield problems. Another benefit is that this process takes advantage of the high degree of accuracy of modern photolithography systems thus enabling more complex devices with multiple functional levels that require alignment. The Orthogonal process uses advanced fluorinated solvents that were designed to be environmentally friendly and recyclable, so little is wasted during substrate processing. Organic electronics can significantly lessen the impact of electronic waste disposal while the company's process can lower toxic emissions during the manufacturing process.

Orthogonal has licensed a portfolio of intellectual property (IP) from Cornell University. This portfolio includes compositions of matter for a suite of photoresists, processes to manufacture the photoresists, and processes to use the photoresist to manufacture organic electronics. A freedom-to-operate search determined that the company would not infringe on any current claims. Internally generated IP will be centered on different formulations of resists. Orthogonal will have a mix of patents and trade secrets to protect the formulations of photoresist and associated solvents.

Business Model and Execution

Organic electronic material manufacturers have not been too successful after several years of effort because there is not a good process to manufacture devices with their materials. Some have approached Orthogonal for potential investment and joint development. The company will continue to collaborate with partners to develop higher-performance, lower-cost photoresists. The go-to-market strategy is to use the marketing and sales channels of their formulation partner and an organic semiconductor industry partner to bring a joint solution (photoresist plus manufacturing process) to display manufacturers that are already customers of these partners. This strategy has already resulted in the production in a pilot-line facility of a flexible e-Reader prototype with an integrated photovoltaic cell in the back. The entire unit can be rolled up, and it does not require batteries to power the display with sufficient illumination. No changes to the equipment were needed to produce this display, only the use of the company's photoresist in the place of the standard photoresist for the patterning of the organic semiconductor (OSC) layer.

The desire to use existing infrastructure influenced the plans of the OSC materials manufacturer - they saw printing as the primary way to make organic electronics a commercial success. They now agree with many of their customers that a photolithographic solution, if demonstrated to be viable, is the best way. Joint work with the OSC manufacturer has demonstrated to both parties that a process with high device yield and performance is possible. The joint roll-out effort represents the optimized process between device and photoresist at this point. Additional benefits from working directly with partners in their pilot-lines is that the company has a better understanding of the areas that need to be improved before the product can be commercially scaled-up. Some improvements are on the process side such as modifications to the wavelength absorption and etch resistance - these can be handled through formula modification. Making polymers on a large scale will require the use of commercially acceptable initiators as well as the elimination of impurities that can act to contaminate the final process. In addition, shelf-life must be improved for final production with the introduction of quenchers. Although this technology is inevitable in the marketplace, it is important that the company executes efficiently to gain market share and establish themselves as the de-facto standard in organic electronics processing.

Financing

Orthogonal has a three-pronged financing approach that includes venture capital, government grants, and investments from strategic partners. The founders and angels invested $0.5M. In a later round, two Canadian super-angel investor groups invested more than a million dollars in the company. The total NSF SBIR funding to Orthogonal in the period 2010-2016 will be approximately $1.8M. There was also additional funding from the State of New York. Product sales, validation from large electronic manufacturers, and the ability to profitably make significant quantities of product, has well-positioned the company to raise venture capital. The CEO has experience raising two rounds of venture capital in the electronics materials' startup Novomer.

Orthogonal has signed a joint development agreement (JDA) with a materials company to produce photoresist for the e-Reader market. This partner is a leading supplier of conventional photoresist for the flat-panel market and has deep expertise in resist formulation and quality assurance. This expertise complements the scale-up process and is the last stage in bringing a product to market. Finally, Orthogonal must have its products be integrated into an e-Reader product line. This will require working closely with a large electronics manufacturer to integrate into their manufacturing processes.

Currently, the company's primary revenue source is product sales - the photoresist solution as well as the developer and stripper solvents. In 2013, one customer used their product in a small e-Reader line that generated additional revenues. If this product is successful, the number of display product lines using its photoresist will increase. This customer and Orthogonal are in the process of optimizing the process to increase yields and reduce costs. There are several companies in the value-chain that would benefit from the company's success thus increasing the likelihood of an acquisition because organic electronic material manufacturers would like to own and control the IP being generated. If they can insure their organic electronic materials are the only materials that can be patterned using photolithography, they will have a much stronger market presence. Additionally, these companies are well-positioned to manufacture the required chemicals. Semiconductor manufacturers will be interested in an acquisition for similar reasons. A manufacturer with the rights to the best process to manufacture organic electronics will have a tremendous advantage. Some of them also have large chemical manufacturing units with the capability to manufacture Orthogonal products. Photoresist manufacturers would also have an interest in acquiring this startup.

NSF Commercialization Assistance and Impact

Orthogonal was introduced to IA in September of 2010 by their NSF Program Director. The startup's R&D efforts aim to enable the large-scale manufacturing of organic and flexible electronic devices by leveraging the existing infrastructure used in the display industry. Currently, proven manufacturing techniques do not exist that can make flexible electronics at scale, and all proposed methods would require the abandonment of billions of dollars of equipment and decades of manufacturing expertise based around photolithography. This R&D effort aims to solve this problem, using photoresist technology that is designed to work with a wide range of materials including organic electronics. Previous research has established the feasibility of these resists, but there is still much work needed to apply the technology to more complex and integrated systems, especially when working with flexible substrates. The company's photoresists will enable electronics manufacturers to produce organic electronics with their existing manufacturing infrastructure and will create a high-performance process capable of making flexible e-Readers with high yield. The startup's needs as identified by IA converged upon the company's R&D and funding efforts. Contemplated issues include potential investor introductions, facilitating partnerships, CEO transition, and building up the management team.

In 2011, the "IA @ CNSE" event sought to connect seven semiconductor-focused companies funded by the NSF SBIR program at the College of Nanoscale Science and Engineering (CNSE) in Albany, New York, a global education, research, development, and technology deployment resource designed to prepare the next generation of scientists and researchers in nanotechnology. The event was divided into two topic areas: flexible substrates and rigid substrates. Orthogonal was featured in the flexible substrate session. IA sought to leverage CNSE's resources and SEMATECH's impressive membership (headquartered in Albany) to accelerate development and help commercialize the technologies being developed by the seven startups. IA activities and outcomes are summarized in the following Table.

IA Activity	Outcomes
Startup Overall Engagement	34 months; 100+ hours of interaction; 175+ e-mails sent to and on behalf of Orthogonal
Building Management Team	Helped with transition to new CEO; helped recruit senior chemist Charles Wright
Operational Assistance	Offered help to previous CEO in different operational areas
Negotiate Commercial Deals	Private Placement Memorandum review and feedback
Introduce Potential Customers/Partners	Samsung; Functional Film Commercialization Center
Find Investors	Southern Capitol Ventures; Dow Ventures; National Innovation Fund; Commonwealth VC; Excell Partners
Identify Domain Experts	Mick Stadler; Brian Johnston; Shelly Weinig; Henry I. Smith
Feature at Trade Shows and Special Events	Featured Orthogonal at the "IA @ CNSE" event; IA sought to leverage CNSE's resources and SEMATECH members to help accelerate and commercialize the company's technology. Featured Orthogonal at the "IA @ EBP: A Materials Science Summit" believing that Orthogonal can benefit from this facility.

In 2012, the "IA @ EBP: A Materials Summit" came to fruition once IA discovered the impressive materials development tools and services as well as high-volume manufacturing equipment housed on the Eastman Business Park (EBP) campus. The EBP materials assets coupled with IA's desire to have U.S. innovations to be manufactured in America made this an excellent opportunity to make NSF SBIR-funded startups aware of this resource and have them connect with local industry and investors. IA alerted the NSF SBIR portfolio to the resource, the event, and then worked with fifteen startups to prepare pitches and posters. As a direct result of the event one local startup submitted a NSF SBIR proposal, one startup (Orthogonal) determined that all of their scale-up manufacturing could be done on-site, another is undergoing diligence by a local fund, and IA is seeking to formalize a relationship with EBP to assist in IA's commercialization assistance efforts.

Where was this startup at the end of federal government funding?

In 2014, Orthogonal, Inc. raised over a million dollars from angel investors – this was matched by NSF, fifty cents to the dollar. Development work shifted because of market dynamics towards organic light-emitting diodes (OLED) display applications. Once the technology was repurposed to OLED patterning, Orthogonal started engaging with display makers to start joint development programs. The process was arduous since multiple iterations and proof-points were needed to convince the display customers that this technology would be viable for mass production, although the advantages of using photolithography to pattern OLED pixels is almost universally recognized. By the end of the project, Orthogonal had started two joint development projects with major Chinese display manufacturers. Both projects will lead to mass production of high-resolution mobile displays if successful.

Orthogonal has released several iterations of the OSCoR photoresist product to customers as development samples over the course of the SBIR program. The current versions of the two product categories are OSCoR 4001 for organic thin-film transistor (OTFT) patterning for flexible displays and OSCoR UP, which is based on the single layer liftoff technology for OLEDs. OSCoR UP is currently being developed jointly with two display manufacturers, with two others in early stages of testing.

The following patents related to this project were filed:

1. WO2016019273
2. WO2016019277
3. WO2016019210
4. US20150030982

Total sales amounted to $360K and contracts worth $500K have been signed. Total revenues from all sources amounted to $2.7M. The number of employees rose from three to eleven during the period of this federal grant.

Orthogonal is primarily engaged with flat panel display manufacturers situated in East Asia (Taiwan, Japan, Korea and China). There are twelve major display manufacturers and several smaller ones. There are approximately eighty display factories in operation. These display manufacturers typically have established businesses making and selling LCD panels to television and cell phone manufacturers for the consumer markets worldwide. Samsung and LG in Korea are at the forefront of building newer, OLED-based displays that have numerous attractive qualities, with Samsung dominating small displays for phones and tablet, while LG makes OLED TVs. Their competitors in other countries are trying to get into display manufacturing, but often lack the capability to manufacture OLED, due to patterning issues. Orthogonal has engaged with both leaders and their competitors with its technology since both would benefit from fundamental improvements in OLED patterning.

Where's the startup now?

Orthogonal has started two joint development projects with Chinese display manufacturers - these projects are designed to take the materials developed at Orthogonal and integrating them into a pilot production line to both make state-of-the-art demos and to work out the process for mass production on larger equipment. Mass production would start in early 2018 if successful. Orthogonal is also engaging with chemical partners to bring the materials to market. Negotiations with Hitachi Chemical are underway to take the materials made by one or more toll manufacturers and formulate and distribute the final product to customers at scale. Orthogonal will not need to build infrastructure to accomplish this scale-up, although expertise in pilot scales will help the company transfer the technology to this partner.

The OLED market is primed for takeoff, with the two Korean leaders making large investments in further OLED technology development and manufacturing facilities ($3B additional investment by Samsung and $9B by LG). New applications such as virtual reality (VR) are emerging as "killer apps" for OLED, and this is particularly good for Orthogonal since VR requires very high resolution, which is not currently possible, but can be made possible with Orthogonal photoresist. The commercial window will not stay open forever since there is intense pressure from the intrenched LCD technology to catch up to OLEDs in performance while driving costs ever lower.

Fine Metal Mask (FMM) technology has reached its limit in resolution, capacity, and yield. Orthogonal leapfrogs limits with ultrahigh resolution, large area patterning. Orthogonal is advancing the field of display manufacturing by lifting the constraints of traditional photolithography. Orthogonal's proprietary photoresist solution allows the direct patterning of a wide range of organic electronic materials for high resolution active-matrix organic light-emitting diode (AMOLED) and flexible display applications.

Case Study Questions

1. Should high-tech products continue to be manufactured in America?
2. Comment on the Lab-to-Manufacture model and IA's role in it
3. How can we leverage other under-utilized manufacturing facilities in America?
4. Discuss the issue of transfer of technology to other nations
5. Why not a purely consumer-driven services economy?
6. Is the business strategy/business model appropriate?
7. What additional skill sets are required in this team?
8. Why raise additional capital – why not grow organically?

9. Identify risk elements (technical, team, market, finance); how to manage/mitigate them?
10. Identify areas where the company needs help
11. Any regulatory issues?
12. Take-away(s)

19   *Measuring What You Cannot See*

Technology Sector:	Instrumentation
Startup:	Anasys Instruments Corp.
Website:	www.anasysinstruments.com
Location:	Santa Barbara, California
Federal Funding Timeframe:	January 2007 – September 2011
Funding Amount:	$1.14M
This narrative is a story of:	A founding team of industry veterans; bootstrap growth; multiple grants; enabling technology; strategic partnering; highly specialized niche market

Company and Team

Anasys Instruments ("Anasys") was founded in 2005 by Kevin Kjoller and Roshan Shetty, senior managers from the Atomic Force Microscope (AFM) industry. Together with key academic scientists the company creates patent-protected products in the field of quantitative nanoscale property measurements based on scanning probe techniques. Anasys designs breakthrough, award-winning products that provide nanoscale probe-based analytical techniques while providing high-quality AFM imaging. The company manufactures atomic force microscopes, thermal scanning microscopes, and nanoscale infrared (IR) metrology equipment. It provides thermal analysis solutions and surface finish-roughness nanoscale measurement tools for the biology, chemical, medical, and semiconductor industries.

The team is made up of the world's leading scientists in the field of nanoscale analysis, and senior business talent with experience in the scientific equipment industry. This combination of technical and market knowledge enables the company to translate its scientific expertise into building a profitable high-growth business. With seventeen years of experience, Kevin is considered one of the world's leading technology and applications experts in AFM and a key player behind the emergence of Digital Instruments/Veeco Instruments ('Veeco') as the leader in this industry. He was most recently the Director of Engineering and Applications at Veeco Instruments where he led a team of forty engineers and scientists with an annual budget of over $10M and was responsible for the engineering of several Veeco products. At Anasys, Kevin is the Vice-President for Research, Engineering and Applications.

Roshan Shetty was the Director of Strategic Investments at Veeco. He was a former investment banker with Alex Brown in San Francisco involved in activities related to Mergers & Acquisitions (M&A) and Initial Public Offerings (IPO) of technology companies. He was also an Operations Manager with KLA-Tencor in the field of semiconductor equipment. At Anasys, Roshan is the CEO and Vice-President of Finance, Sales, Manufacturing, and Administration.

The company introduced the nano-TA in 2006 that launched the field of nanoscale thermal property measurement. In 2010, Anasys introduced the nanoIR platform to develop the field of nanoscale infrared measurement. In 2012, the firm pioneered the field of wideband nanoscale dynamic mechanical spectroscopy. The company introduced nanoIR2, the second generation of their nanoscale IR spectroscopy platform, a breakthrough system that features top-side illumination to greatly expand the range of samples that can be studied. It combines the nanoscale spatial resolution capabilities of AFM with infrared spectroscopy's ability to characterize and identify chemical species. It is a powerful, easy-to-use multifunctional platform that provides full-featured atomic force microscopy and nanoscale thermal and mechanical analysis. Since its founding, Anasys has developed and introduced multiple major award-winning products such as the AFM+ thermal analysis product that was a 2008 R&D100 Award winner. The first-generation nanoIR won the 2010 R&D100 as well as the Microscopy Today Innovation Award.

Anasys has assembled a distinguished set of scientific advisors to guide it as it revolutionizes the field of nanoscale IR spectroscopy and nanoscale thermal analysis measurements. These advisors are international thought-leaders in academia and industry with expertise in polymers, IR spectroscopy, mid-IR sources, semiconductors, biomaterials, thermal probes, thermal analysis, and scanning thermal microscopy. They include Professor Bill King, University of Illinois, a scientific co-founder of Anasys and the world's leading authority on nanoscale thermal probes and property measurements, Professor Alexandre Dazzi, University of Paris-Sud, the inventor of the Photo-Thermal Induced Resonance (PTIR) technique, the basic technology behind the company's R&D efforts, and the only proven technique to obtain sub-100 nanometer IR spectroscopy and imaging, and Dr. Ken Babcock, CEO of Affinity Biosensors and former General Manager of Veeco's $80M a year AFM business. Ken is also an investor in Anasys.

Market Opportunity

IR spectroscopy led to the discovery of synthetic rubber during World War II. Since then, it has remained a critical and ubiquitous analytical technique which is by itself a $1B per year industry. However, its spatial resolution limitation has seriously hampered researchers in the pharmaceutical and strategic materials industries where multibillion dollar nanotechnology investments have led to a massive need for information on nanoscale chemical composition. As reiterated by endorsements from companies such as Dow Chemical, Dupont, and Eastman Chemical, the enormous impact of nanoscale infrared measurements ranges from new materials discovery, interfacial property improvements in high-value applications like automobiles, and breakthroughs in the life sciences in general and particularly in early prostate cancer screening. This will have a disruptive impact on the $250M per year AFM industry.

The AFM is a workhorse instrument for nanotechnology research and the information it provides is a key driver of the $1.2B National Nanotechnology Initiative (NNI). However, university and federal lab researchers insist that the most serious bottleneck facing this user community is the AFM's inability to provide chemical analysis functionality. Anasys is developing the world's first technology platform for sub-100 nanometer (one-billionth of a meter) IR spectroscopy and imaging, a 50-times improvement in resolution over the current state-of-the-art. This platform is based on the proprietary and patent pending PTIR technique that will shatter the optical diffraction limit that has plagued IR spectroscopy and imaging for the last sixty years. Satisfying these unmet needs in nanoscale thermal and chemical measurements will enhance our nation's nano-manufacturing infrastructure.

The initial target market is the chemical and materials industry. Four market segments have been identified:

AFM Users:

The Atomic Force Microscope is the leading tool for nanotechnology research. It is versatile and can be used for all types of nanoscale measurements. The current AFM market is growing at 15% a year with annual unit sales of a thousand with an installed base of seven thousand. However, the most important limitation of this tool is its inability to measure the chemical properties of a sample.

IR Microscopy Users:

This user group consists of R&D scientists in the polymer and life sciences industry who seek localized chemical information from IR spectra of their samples. The IR microscope market is currently $82 M/year and growing at 7% annually. The most important limitation of this tool is its inability to resolve features of below ten micro-meters (one-millionth of a meter).

IR Spectroscopy Users:

IR spectroscopy is a common technique employed in industry for chemical identification used in applications ranging from quality control to R&D. The total size of this market is $1B/year and growing at 4.5% annually. The most important limitation of this tool is its inability to work on samples smaller than a few millimeters across. The need for smaller samples is most felt by users from the organic materials, forensics, and life sciences industries. Users range from technicians to R&D scientists who use it to solve a specific problem and appreciate its ease-of-use. An estimate of the total available market for a small-sample IR platform is 20% of the overall market which translates to $200M/year. The annual growth rate is 10%.

Advanced Optical Microscopy Users:

This user group comprises R&D scientists in the polymer and life sciences industry whose microscopy applications require high resolution and who value additional information on their samples. The total market size is $200M/year and growing at 10% annually. An important limitation of this tool is its current inability to provide chemical information on samples. IR microscopes are not used since its resolution of 10 micro-meters is far worse than the optical imaging resolution of 1.5 micro-meters.

Technology and Product

The National Science Foundation (NSF) Small Business Innovation Research (SBIR) funded project seeks to develop the prototype of a characterization system which can perform IR spectroscopy and imaging at sub-100 nanometer (nm) spatial resolution and thus break the 5-micron resolution barrier that has limited IR spectroscopy for the last fifty years[25]. This fifty-times breakthrough in the degree of spatial resolution is enabled by the company's proprietary PTIR technique whose feasibility has already been demonstrated in prior work funded by the same NSF SBIR program. The total NSF SBIR funding to Anasys in the period 2009-2014 has been approximately $2.2M. The prototype is based on a broadly tunable mid-IR laser invented by one of the company's scientific co-founders. Research efforts now focus on fully characterizing and optimizing the PTIR technique which in turn involves optimization of the nanoscale probe technology, IR illumination, detection algorithms, and the minimization of non-IR absorption related contrast. This is a multi-disciplinary effort that will leverage the team's cutting-edge competencies in IR physics, electro-mechanics, optics, and software.

The previous NSF SBIR-funded R&D project enabled commercialization of nanoscale Dynamic Mechanical Analysis (DMA). Conventional DMA works by applying an oscillating stress to a sample and measuring the time-dependent strain. Analysis of DMA data gives information, for example, about material stiffness, viscosity, thermal transitions, and activation energies. DMA is a critical and widely used tool to measure the viscoelastic properties of bulk materials, but it suffers from three key limitations: slow speed, limited frequency range, and the lack of spatially resolved information. Large and growing material classes employ nanoscale composite structures to achieve desired material properties. No current tool can rapidly examine the temperature-dependent viscoelastic response of these materials at the scales they are being engineered. To address this unmet need, the company is developing instrumentation based on atomic force microscopy (AFM) using rapidly heated AFM cantilever probes. Specifically, this nanoscale platform will provide variable temperature DMA in seconds; measurement frequencies three orders of magnitude higher than conventional DMA; and spatial resolution down to less than 100 nm.

The nanoscale ultrafast dynamic mechanic analysis (nu-DMA) product addresses a market need for reproducible discrimination, identification, and characterization of components in complex materials. It does this by enabling rapid, high-frequency and ultra-sensitive measurements of glass transitions and other viscoelastic properties of commercially important polymers, including highly filled, cross-linked and/or crystalline polymers and thin films. Initial orders valued at over $1M for the preliminary product from university and industrial researchers provides validation of possible far-reaching impacts in critical enabling systems such as block co-polymers, impact modifiers, and polymer blends used in several multi-billion industries ranging from chemicals, medical devices, electronics, and automobiles. This product will appeal to customers that the company is currently addressing with their nanoscale thermal and IR technologies creating beneficial sales and marketing synergies. The estimated addressable market is $50M/year.

The current NSF SBIR-funded R&D is focused on developing new forms of atomic force microscope-based infrared spectroscopy (AFM-IR). It will provide chemical spectroscopy and compositional mapping with extremely high spatial resolution. The product resulting from initial research has already led to orders valued at half-a-million dollars from a major international chemical company and a major gas-and-oil services company. The company expects this product to be an approximately $30M/year market within five years. The IP strategy consists of protection by a suite of twenty worldwide potential patents, five of which have been filed, in addition to which the company has exclusive licenses to fifteen others from TA Instruments.

Business Model and Execution

The Anasys business model is based on continually investing in its current core competencies of developing new techniques for nanoscale property measurement, engineering product development, and customer-focused applications development. All aspects of manufacturing including subassembly work are outsourced. The final assembly and quality control prior to shipment is done in-house. Competition is expected from the leaders in the four market segments identified above but the company believes its products deliver superior performance in measurement throughput, the applicable range of temperatures and frequencies, spatial resolution, ease-of-use, and the range of measurable materials. There are other competing technologies like confocal Raman and infrared microscopy being used in mature products from larger spectroscopy companies, though with much coarser spatial resolution than these new products from Anasys.

The key commercialization team members consisting of Roshan, Kevin, Dave Voci and Craig Prater have deep, relevant industrial experience. All four previously worked at the AFM leader, Digital Instruments/Veeco, recently acquired by Bruker Corporation. Additionally, Roshan has strategic partnership experience from investment banking and Dave Voci was responsible for business development of the helium ion microscope at Carl Zeiss which he grew from $0 to $15M/year over four years. Prior to this, Dave was responsible for a $4M/year sales territory at Digital Instruments' AFM division. Craig was the former Chief Technologist at Digital Instruments/Veeco.

The company uses distributors to sell their products worldwide and are transitioning to a direct salesforce in key markets - U.S., Japan, and Western Europe. They employ the direct-sales model with their current product in the nanoscale thermal analysis arena. The initial customers for their nanoscale IR products are the same users of their current nanoscale thermal products – this makes for an easier adoption process. Their strategy is to launch initial products (both thermal and infrared) to the AFM industry to obtain customer validation and feedback required to make the technique robust and easy-to-use. Using this feedback, they plan to launch the next generation easy-to-use tool targeting the bulk-analysis industry. They have one direct salesperson operating a network of four distributors to cover different U.S. regions. They plan to shortly increase the number of such personnel. They have a strong distribution channel in Japan and Korea with dedicated distributors who have invested in their products. The company has recently started to invest in China and India with one dedicated person in each country. Their sales have historically been weaker in Europe - to correct this anomaly the company has recently appointed three new distributors covering major European countries.

Current financing consists of cash-flow from product revenues, and bank financing to smooth over monthly cash-flow related issues. Beyond the initial funding requirements for new product launches and associated higher R&D and marketing costs, their financing model trends towards R&D and marketing costs as a percentage of revenues that are slightly more modest than the historical values for Digital Instruments, a pioneering company in the field of AFM.

NSF Commercialization Assistance and Impact

Anasys was introduced to Innovation Accelerator (IA) in 2009 by their NSF SBIR Program Director. There was mutual interest in identifying and prioritizing key issues, especially those that aligned with the resources available to IA. An in-depth needs analysis was conducted. The startup's needs as identified by IA converged upon the company's commercialization efforts. Contemplated issues included the belief by the company that it would first need to launch an accessory device for the smaller atomic force microscopy (AFM) market before they could complete their standalone device. Soon thereafter, they determined that Anasys could launch both such devices in 2010 provided they had a close partner in Veeco, the leader in the AFM market.

Was there any downside to partnering with Veeco for the accessory device in the AFM market, where the accessory device would be an addition to Veeco's AFM machines while launching their own standalone machine for the infrared instruments market? Another key issue was the need for a strategic investment with the right-of-first refusal. A corresponding term-sheet was required from Veeco. IA introduced Marco Rubin and Marcus Ruark, experienced term-sheet negotiators, to the management team. IA conducted outreach and background diligence to provide potential IA Advisors to the company. IA monitored the engagement for compatibility and to identify emergent areas of potential assistance.

In January 2011, as previously indicated, the inaugural "IA @" event series was held in conjunction with the "Nuts & Bolts" entrepreneurial course at MIT. Three NSF SBIR-funded startups were selected to present as innovation 'case studies', Anasys being one of them.

Testimonial

"Thank you for introducing Anasys to IA - after we passed early vetting, we were put in touch with two mentors, Marcus Ruark and Marco Rubin. One has a corporate VC background, and the other has founded companies/raised money/exited et cetera. The immediate issue facing us at Anasys was whether we should partner with a larger company in launching our product or go to it alone. Our technology breakthrough is in nanoscale IR spectroscopy which positioned us between two instrumentation sectors, the $1B/year IR spectroscopy sector and the $250M/year AFM sector. Most of the key folks at Anasys were senior executives at Veeco, the leader in the AFM industry. We are constantly in touch with Veeco and have an OEM deal on our prior nanoscale thermal analysis technology with them."

"Marcus and Marco spent time with us to sort through the pros and cons of a partnership, the timing of a partnership and its implications on company valuation. I found those discussions to be crucial in clarifying issues for us. I have a prior investment banking background and could appreciate the value that Marco and Markus's input provided. We eventually decided to go it alone and launched our product, called the nanoIR at Pittcon in March 2010. While still early to say, we at Anasys are convinced that it was the right decision given the strong interest we've seen so far in the market. We have not precluded a partnership with Veeco or a leader in the IR Spectroscopy field such as Perkin Elmer but if we do end up going that road, we'd be in a much stronger position to negotiate terms in our best interest."

"I'd like to applaud and thank the NSF (and IA) for having set up the IA program to provide SBIR companies with such high-level expertise (at zero cost to us) in sorting out complex decisions that have long-term implications for a company's success."

Where was this startup at the end of federal government funding?

The commercialization effort was begun with an extensive analysis of the requirements of the customers utilizing a process called "Voice of the Customer". This exercise consists of asking customers open-ended questions with the goal of identifying their problems/needs and the value/priority of those needs. Besides confirming the company's premise that nanoscale IR will solve important problems for both AFM and IR users, it also helped understand the key requirements (and priorities) of these customers regarding spatial resolution, spectra interpretability and wavelength range of interest. Anasys announced the product at Pittcon in March 2010. Since Pittcon has over two thousand exhibitor booths and over fifteen thousand attendees, the firm decided on a targeted pre-show marketing effort to get its potential customer base excited and to seek Anasys out at Pittcon. Anasys' CTO's technical talk on nanoIR at Pittcon was a standing-room only event, drawing significantly more attendees than the previous talks in the session.

Sources of revenue include contracts totaling approximately $2M from potential industry partners, federal agencies, and university research teams. A Fortune 500 customer was originally focused on purchasing a standard AFM and contacted Anasys on hearing about the nanoIR. During the first conversation they stated, "We are a multibillion-dollar company and so, prefer to only deal with large companies." Hence, the startup approached two other Fortune 100 companies. They stated – "We are approaching you as the nanoIR product seems unique and we'd like to see if we get any interesting results." They sent Anasys samples and were pleasantly shocked when the firm correctly identified the sub-micron sample constituents via the nanoIR. Potential customers for the company's products are in the U.S., Europe, and Japan.

The Table below summarizes IA activities and outcomes related to Anasys.

IA Activity	Outcomes
Startup Overall Engagement	48 months; 150+ hours of interaction; 120+ e-mails sent to and on behalf of Anasys
Recruit/Assist Management Team	Assistance was provided to further engage CEO and Board members, and to help the CEO in all operational areas.
Negotiate Commercial Deals	Helped structure and negotiate commercial deals
Introduce Potential Customers/Partners	Intel, InSituTec, and SEMATECH
Find Investors	National Innovation Fund, Commonwealth Ventures, and Siemens
Identify Domain Experts	Connected the team to domain experts Marco Rubin, Marcus Ruark, Richard Kivel, and Bryan Rice
Featured at "IA @MIT" event	

Where's the startup now?

The Anasys founding team is made up of the world's leading scientists in the field of nanoscale analysis, and senior business talent with experience in the scientific equipment industry. This combination of scientific and market knowledge enables Anasys to translate its revolutionary scientific expertise into the building of a profitable high-growth business. Anasys is the world leader in nanoscale IR spectroscopy dedicated to delivering innovative AFM-based nano-spectroscopy products and solutions that measure spatially varying physical and chemical properties with nanoscale spatial resolution in a diverse range of fields, including polymers, 2D materials, materials science, life science and micro-electronics industry. The team is focused on providing robust nanoscale chemical spectroscopy and analysis. With a researcher's productivity always in mind, the company delivers integrated hardware and software solutions that clear the path to the user's next discovery. Nanoscale IR spectroscopy is complemented with techniques such as mechanical, thermal, and electrical analysis to provide unique multi-modal capabilities.

Anasys develops, manufactures, and markets products and solutions that measure nanoscale material properties. It offers nanoIR2, an

AFM- based IR spectroscopy platform; and nanoIR, a probe-based measurement tool that reveals the chemical composition of samples at the nanoscale, as well as combines IR spectroscopy and AFM enabling the acquisition of IR spectra at spatial resolutions. The company also provides AFMs; nano-TA, a nanoscale thermal analysis module that allows sub-100nm local thermal analysis for commercially available AFMs; and SThM, a scanning thermal microscopy module, which include controllers, software, and probes enabling AFMs to do temperature mapping of their samples with 0.1°C resolution. In addition, it offers ThermaLever probes, a batch fabricated silicon probes composed of doped silicon; and SThM probes that allows measurements of temperature of the apex of the probe. The company's products are used in a range of fields, including polymers, materials science, life science, semiconductors, data storage, biology, chemical, medical, and other applications.

Anasys was founded in 2005 and is based in Santa Barbara, California with an additional office in Mannheim, Germany.

Case Study Questions

1. Critique the value proposition and the business strategy/business model adopted
2. The role of NSF SBIR funding and its constraints
3. Identify risk elements - technical, team, market, finance – and how to manage/mitigate them?
4. What are the barriers to entry for Anasys, and its competitors?
5. Identify areas where the startup would need help
6. Conduct a start-up valuation exercise
7. Comment on the growth strategy, and exit strategy should there be one
8. Take-away(s)

20   Plastic Baby Bottles Can Be Safe

Technology Sector:	Life Sciences
Startup:	Plastipure, Inc.
Website:	http://plastipure.com
Location:	Austin, TX
Federal Funding Timeframe:	July 2010 – December 2014
Funding Amount:	$0.78M
This narrative is a story of:	A public health issue; David v. Goliath – struggling small company v. powerful entrenched interests and incumbent players; three startups as one; positive press coverage, lawsuits, and unfavorable judgments; strategic partners; licensing model; scale-up/production issues

Company and Team

PlastiPure was founded in 2000 by University of Texas neurobiology professor Dr. George Bittner with the mission of creating plastic bottles free of all chemicals with estrogenic activity (EA). Such chemicals have been linked to serious health issues, from altered behaviors to greater rates of some cancers. The company's products directly address growing scientific and public concerns about the safety of commercial plastics that release endocrine disrupting chemicals. Over a decade, Bittner, and his team of experts in cellular and molecular biology, endocrine physiology, and polymer chemistry developed the technology and created intellectual property. In 2005, PlastiPure's sister company CertiChem developed early models and advanced testing techniques to predict and detect EA. The firm leveraged these capabilities over the following three years to develop multiple patents; an improved computer simulation model; a large EA database on chemicals, compounds, additives, materials, processing aids, and products; and extensive EA material and product remediation knowhow.

In 2008, it became the first company to introduce a line of hundred percent EA-free plastic bottles using a large portfolio of resins, colorants, additives, and processing aids along with approved conversion processes. In 2009, the firm released the world's first certified EA-free reusable water bottle manufactured through a partnership with Hydrapak. That same year the company developed custom EA-free materials for a few customers and signed a licensing agreement with an advanced polymer manufacturer. In 2010, it developed EA-free technologies for flexible packaging thus allowing future production of film applications. Also in 2010, to demonstrate viability, PlastiPure signed additional agreements with two other companies to deliver simple EA-free prototype bags to these customers. It is now engaged with partners for the manufacture of additional EA-free products such as baby bottles, food packages, and medical devices.

PlastiPure's core team of biologists, polymer chemists, industry specialists, and business and sales personnel offer a diverse combination of expertise and experience to match the company's vision and technology. It employs key scientific and business staff internally, taps recognized functional experts as consultants, and partners with crucial supply-chain vendors such as resin manufacturers, processors, distributors, and consumers.

Michael Usey, CEO, has over twenty years of experience in successfully growing companies specializing in the commercialization of breakthrough technologies. He has consummated key industry and retail partnerships licensing the company's EA-free patents for use in a variety of infant feeding, food packaging, beverage, and personal care products. Prior to this, Michael was CEO of an unprofitable business that he turned into the software leader for multi-location optical retailers. He earned a B.S. in bioengineering and an M.S. in electrical engineering at Texas A&M University.

George Bittner is the founder and Chief Scientific Officer of Plastipure, and the CEO of CertiChem. With more than one hundred and fifty peer-reviewed publications in molecular biology, electrophysiology, synaptic biophysics, and the biochemistry of nerve regeneration, Dr. Bittner was elected as a fellow of the American Association for the Advancement of Science in 1992. George earned his B.A. in chemistry at Duke University and his Ph.D. at Stanford University.

Stephen Suknaic, Sales Manager, has over two decades of specialized sales and marketing experience in the polymer chemical industry. Previously, Steve with his extensive expertise in supply-chain management managed international business for a Fortune 500 company. He earned his B.S. and M.B.A. degrees at Florida State University.

Chun Zhi Yang is a scientific consultant to Plastipure and the Chief Operating Officer of CertiChem. Dr. Yang was previously the research director of biotechnology at Monsanto. She earned her M.D. at Hunan Medical School, and her Ph.D. at the Pasteur Institute in Paris.

Stuart Yaniger, R&D Vice-President, has over twenty-five years of experience in plastics, product development and scale-up. He previously took two companies from idea to successful high-volume production. After obtaining his Ph.D. in polymer science from the University of Texas, Stuart was a postdoc at the University of Pennsylvania with Nobel Laureate Alan MacDiarmid.

Andy Starr, Vice-President for Business Development, has an MBA from UCLA and significant experience commercializing new polymer and biotech products. He was formerly President of Neocork Technologies.

Daniel Klein, Senior Research Scientist, obtained his Ph.D. in polymer science from the University of Akron. He has over ten years of experience in polymer R&D and product development, with particular expertise in synthetic organic chemistry.

Market Opportunity

Consumers, the popular press, nongovernmental organizations, and legislators are increasingly concerned about the safety of plastics almost all of which release chemicals with estrogenic activity. These chemicals are contained in and leach from many common consumable, packaging, and retail products. While estrogens occur naturally in the body, many scientific studies have reported significant health problems when synthetic chemicals are ingested that mimic or block the actions of natural hormones. These potentially harmful products are widely present in infant feeding products, food storage and preparation bags, cling/stretch wraps, lab and pet feed bags, and medical packaging. Health-related problems from EA include early puberty in females, obesity, and increased rates of breast, ovarian, testicular, and prostate cancers. It also leads to reduced sperm counts in males, altered functions of reproductive organs, and altered behaviors and learning rates.

PlastiPure's research has determined that between five-to-twelve thousand chemicals are thought to have EA chemicals introduced throughout the plastic manufacturing process, including polymer synthesis, resin development, processing, and final product finish. Consumers are increasingly aware that the underlying cause of the harm of Bisphenol-A (BPA) and other chemicals that imitate estrogens is their estrogenic activity. Their concerns will drive rapid demand for comprehensively safer products throughout multiple industries. Many markets have already been disrupted by introduction of perceived healthier products. This is especially true in the infant feeding market, where initial concerns about BPA drove the nearly total replacement of plastics commonly used for baby bottles with non-BPA-containing alternatives within the space of only a few years. Consumer demand and the potential for regulation and litigation have extended that trend to include nearly all retail plastic items intended for infants and children. This disruption will force safer no-EA products to become widely available, with the infant/children space being particularly receptive. Usually, non-plastic solutions chosen by consumers are more expensive and have less appropriate fitness-for-use characteristics. Additionally, most BPA-free plastics have been recently demonstrated to be no safer than BPA-containing plastics, as most leach other chemicals with easily detectable EA.

The $18B flexible plastic packaging represents the largest segment of the plastics packaging market of which approximately sixty percent is used in the food and beverage market. This market is extremely consumer-driven due to high customer awareness of packaging dangers and available options. The company has chosen to target the segments with the highest potential health impact and consumer visibility - infants and children. It is estimated that twenty million baby bottles are sold annually in the United States alone. Consumers are extremely interested in the safest baby bottles available, as evidenced by the growing market penetration of glass and stainless-steel bottles. Success in these market segments will likely cause EA-free technology to cross-over into other, even larger areas such as general food packaging, stand-up pouches, can linings, flatware, and any other plastic product which directly contact foods or beverages.

PlastiPure faces competition from companies focused on producing "safer" plastic products such as resin manufacturers currently producing materials that are perceived to be safer, but none are licensed to develop EA-free plastics. The company has exclusive patents on these methods and formulations. It directly challenges its competition by providing a variety of products that are competitively priced and available through existing supply channels. The firm has, in small volumes, the first EA-free wholesale and retail products on the market today - baby bottle liners, cling/stretch wraps, food storage bags, and food packaging and preparation bags. One end-user that has already demonstrated high consumer demand for such products is partnering with this startup to commercialize and distribute these products once they have been sufficiently developed.

Early adopters of its EA-free technology also have significant branding and positive press coverage as companies focused on the health and safety of their customers, and as environmental stewards. These companies also mitigate the growing potential liability that many organizations are facing due to consumer, legislative, and scientific concerns on issues associated with endocrine disruption, especially estrogenic activity. Public pressure is mounting to ban EA-inducing products through public health warnings, and with legislation being proposed and passed in many countries. Laws and regulations intended to solve the estrogenic activity problem by banning one chemical at a time will not lead to a comprehensive solution. The problem is too pervasive, and producers of conventional plastics offer no alternatives. Manufacturers who have successfully responded to singular chemical bans could be significantly hurt when systemic change in the plastic channel is driven by newly proposed legislation or growing press coverage of endocrine disruption. Conversely, these companies can pick up significant market share if they proactively take advantage of the easily implemented technology that PlastiPure has created to make EA-free materials and finished goods.

Technology and Product

BPA is used to manufacture polycarbonate plastics for food and beverage containers, and epoxy resins typically found in dental sealants and as food contact liners in cans. Phthalates are a class of chemicals used to soften plastics and make them more flexible. They are consistently found in children's plastic toys. BPA and phthalates have been highly publicized for their EA, but they're only two of hundreds of such chemicals commonly used to produce plastics. Some companies promote their efforts to eliminate BPA and phthalates from their products while leaving in other chemicals with dangerous levels of EA. Moreover, BPA-free, or phthalate-free products are not usually EA-free. Sometimes the chemicals substituted for BPA or phthalates show more EA than the original chemicals. Replacing individual chemicals in plastics is neither an efficient nor effective way to solve health problems caused by chemicals with EA.

The plastic industry follows a set of guidelines mandated by federal and state laws, best manufacturing practices, and reasonable business policies. With the increasing concerns that scientists, consumers, and now legislators have about endocrine disruption, these rules are rapidly changing. Companies that do not respond quickly and proactively are likely to lose market share, face class-action lawsuits, or be locked out of the market through exclusive agreements and patent protection around EA-free technologies. The introduction of PlastiPure's patented breakthrough EA-free technology that addresses estrogenic chemicals at all levels of the supply-chain, polymer manufacturers, processors, retail product companies, and consumers have a viable choice for safe plastic materials and products[26]. The company's technology is meant to improve, not replace, current plastics. Its production methods are easily implemented without additional tooling costs. The firm uses the most sensitive available EA assay tests to ensure that its products are the safest. It is the first and only company producing plastics, silicones, paper, and other materials that can be recycled and certified free of estrogenic activity.

PlastiPure has knowhow and multiple patents on all known practical methods to make desirable EA-free materials and products. This IP portfolio has been built using its improved computer simulation models using years of empirical data collected from CertiChem, its strategic partner, that has developed novel robotic assays to determine the presence of EA rapidly and accurately These assays are being evaluated by federal agencies such as NIH and EPA. The startup also has access to a proprietary EA database which contains the results of many thousands of tests on chemicals, compounds, materials, processing aids, and finished goods. This combination of knowhow and EA data has allowed development of commercially viable materials and production methods to inexpensively produce finished EA-free products using existing manufacturing and process tooling. The firm has the necessary lab-scale processing and testing equipment to develop new formulations to process into a variety of EA-free product prototypes.

Utilizing resins developed from NSF SBIR funding, PlastiPure showed that it can produce EA-free films with properties that are comparable to commercial films and bags, and that it can produce a variety of EA-free polymer flexible packaging formulations to process into EA-free films. Discussions with distributors and product companies in the infant and child feeding market helped to carefully select applications based on production and commercial viability. The most compelling commercial applications are also the ones that have the highest societal benefit: products for pregnant mothers, infants, and young children. The startup believes that single-use flexible packaging will produce the highest value to customers for the least risk and expense. Any additional engineering and cost requirements can be overcome by leveraging its strong plastic-industry relationships, design experience, and equipment that mirror those being used for current mass-produced flexible packaging.

Plastipure offers resins and formulations for EA-free paper and silicone-based products; applications for plastics processing and manufacturing; testing services for validating hormonal activity; and product development and certification. Due to the commercial demand for its products, it was necessary to find supporting partners in the markets for resin supply/distributing, packaging, raw materials, manufacturing, and distribution of liners, films, and bags. Manufactured with the same capital equipment and similar processes as conventional plastics, its resins usually cost just a few cents more to produce with only pennies added to the price of the final product. The firm manufactures certified EA-free wholesale bottles for personal care by working with a variety of molders to deliver both stock and custom packaging for the wholesale market.

In the group of products spanning baby, personal care, fitness, food packaging, and medical devices, the three most important products are baby bottles, sippy cups, and pacifiers. The company has requests from retailers, product manufacturers, and consumers to get these key products into stores. The firm has a licensing agreement with a partner that markets a suite of brands in this market segment. It selected a sippy cup as the product most likely to be quickly delivered and successfully marketed. It is a product that this partner has the infrastructure to rapidly launch and scale. The partner will financially support this product if a prototype can be built that meets both EA-free and commercial standards.

Partner companies will not engage in early R&D - they require working prototypes that are commercially viable for large-scale production and remain EA-free even after common-use stresses such as microwave radiation, sunlight, or heat. While this partner is not ready to fully commit funding for a production cup until a prototype can be built which proves out this EA-free technology in a commercially viable production model, it is willing to partially defray initial prototyping costs through assistance in design and part processing.

Business Model and Execution

The roadmap to substantially open EA-free markets has been demonstrated by the previous disruptive uptake of BPA-free products. PlastiPure is pursuing a business model since regulation is not a prerequisite for rapid market change, but the threat of later regulation/litigation is helpful. Widespread awareness of the issue, evidence of negative health impacts must be defensible but not required to be definitive, and solutions must be implementable within the existing supply-chain for competitive pricing such as no retooling requirements and onerous processing changes. The company uses a multi-tiered business model to maximize uptake of its patented technologies across many industries. The revenue model has three components: Produce breakthrough EA-free products that are most likely to have a large health benefit to consumers and increase consumer awareness of EA-free and the PlastiPure brand; license co-developed EA-free retail products to wholesalers, retailers, and product manufacturers; and license EA-free materials to manufacturers to drive consumption of EA-free products in multiple market segments.

Revenues could also be garnered with associated products such as baby bottle nipples and pacifiers. PlastiPure also provides the most sensitive testing assays available to detect EA and has a large database of chemicals, additives, antioxidants, colorants, and other materials that can be quickly substituted for reformulations to produce EA-free materials. In most cases, changes can be easily and inexpensively implemented either to remediate an existing product or to launch a new line. The firm has a wide variety of production-ready EA-free materials at competitive prices that have the same processing and physical characteristics as conventional, EA-containing plastics. It also has processing aids, purge agents, mold releases, colorants, and other materials and supplies that simplify processing an EA-free material. The company can work within a client's existing supply-chain or provide its own materials and processing.

While PlastiPure has developed the underlying technologies, it needs customers with operating capital to aggressively develop, test, launch, and advertise these key products. The company has licensing agreements and is in active negotiations with companies that want to brand and sell these products that are commercially ready. Negotiations are more likely to be productive if the company can deliver prototypes of flexible packaging applications that pass fitness-for-use and stringent EA tests. Its preferred commercialization path is to charge development costs to take its existing technology to product readiness. This is the early cost that product manufacturers have been unwilling to pay; NSF SBIR funding helped to overcome this barrier. The total NSF SBIR funding to the startup in the period 2010-2014 has been approximately $780K. Overall funding from federal agencies and investment from the private sector together total $3.5M. After products are commercially ready, the firm will charge for ongoing testing required to maintain EA-free status, plus a royalty based on products sold.

An alternative revenue model for flexible packaging is for PlastiPure to receive significant private funding and retailer interest. In this model, the company would wholesale its EA-free flexible packaging products directly to product manufacturers or retailers and use private equity to provide working capital and funds for inventory, marketing, and promotion. The company has strong supply-chain relationships to make this model work but would prefer not to enter the more capital-intensive wholesale/products business. This model could be more profitable initially than the royalty-only model but also carries more risk. If the firm did not have strong partners to bring EA-free products to market, this wholesale model would be the most appealing. Either strategy benefits from the company not only providing the materials, processing methods, and certification for EA-free but also championing the EA-free advantages to consumers through a variety of channels.

PlastiPure anticipates competition from resin manufacturers currently producing products that are perceived as safer (e.g., BPA-free). Large chemical companies are not likely to operate in the EA-free market for several years until consumer demand and litigation/regulatory threats force them to do so because of supply-chain/inventory inefficiencies and potential liability. By the time they enter the market, the firm and its licensed partners hope to be the established brand leaders. Following corporate decisions to go EA-free, it is more likely that these large companies will choose to work with or acquire the company rather than waste several years and many millions of dollars to try to duplicate the technologies developed by PlastiPure and CertiChem. The company will address market threats by preempting or co-opting much of this competition by leveraging relationships with its partners or resorting to litigation if required, and by relying on protection offered from five issued patents.

For normal operations, the company will depend on its growing revenue stream from existing products such as EA-free water bottles, baby bottles and liners, cling/stretch wraps, and a variety of bags. Customers are product manufacturers and retailers. While investors have shown significant interest in the company, venture capital firms would prefer to see demonstrated products on the market. A more likely scenario is faster revenue growth beginning in 2014 driven by safer products for infants and young children, and cross-industry demand for multiple products as demonstrated by demand for BPA-free products years earlier. Revenue projections are based on existing products such as water bottles and personal care bottles, and services such as material development, testing, and consulting, and later products such as baby bottles, baby bottle liners, cling wraps, bags, and other flexible packaging. Additional revenues are generated by licensing fees charged to material and product companies.

NSF Commercialization Assistance and Impact

An NSF/SBIR Program Director recommended PlastiPure to IA. This was a strong endorsement of the company because this Program Director customarily only recommends a small but impactful set from his portfolio of startups to IA to support. The main challenge facing PlastiPure was that their market direction and market segment were not fully clarified.

IA arranged for PlastiPure to pitch directly to a senior Director, Corporate Office of Science and Technology, at a Fortune 30 U.S. multinational; IA, at the startup's recommendation, also pitched to this multinational on the use of its solution in baby bottles, only to discover that the multinational company was no longer in this business. The company followed up with direct discussions with the multinational with IA being available to assist in these discussions as deemed necessary.

IA met with a major, U.S.-based manufacturer of cosmetics to present to them the company's solution. The manufacturer validated the value proposition but also indicated that it would be the manufacturer's plastics suppliers, rather than the manufacturer itself, that would be the end-user of the company's solution. IA worked with the manufacturer to approach these suppliers on the startup's behalf.

IA met with a leading global manufacturer of chemicals and plastics. Plastipure then had direct discussions with this manufacturer. The company is now reviewing potential business opportunities using the firm's technologies.

"IA @ EBP: A Materials Summit" - IA believed that the startup could benefit from the manufacturing resources available at this facility. The company was one of four NSF SBIR/IA case studies featured at Stanford University.

In July 2011, IA worked with Plastipure so that it was one of four featured startups at the "IA @ Stanford" event. In 2012, IA again worked with Plastipure so that it was one of fifteen startup companies featured at the "IA @ EBP: A Materials Summit" event. In general, IA facilitated market feedback and traction to Plastipure through several introductions to potential customers and partners as summarized in the previous Table.

Where was this startup at the end of federal government funding?

PlastiPure began to receive significant attention from the press in 2011, after it published test data for hormonal leaching of common plastic materials and consumer products. With this publication, PlastiPure received many interviews and positive press coverage. Testing sales began to increase, and investors became interested in the company. Unfortunately, shortly thereafter, Bittner's startling results set off a bitter fight with the $375-billion-a-year plastics industry. The American Chemistry Council, which lobbies for plastics makers and has sought to refute the science linking BPA to health problems, has teamed up with Tennessee-based Eastman Chemical, the maker of Tritan, a widely-used plastic marketed as being free of estrogenic activity, in a campaign to discredit Bittner and his research. Eastman's offensive is just the latest in a wide-ranging industry campaign to cast doubt on the potential dangers of plastics in food containers, packaging, and toys.

A month after Bittner's study appeared, the American Chemistry Council and the Society of the Plastics Industry argued that CertiChem's findings were "unconvincing"; they maintained that just because a substance behaved like estrogen in a culture dish did not mean it would do so in animals or humans. Once its own data had been published, Eastman set out to bury Bittner's findings. In August 2012, Eastman Chemical sued CertiChem and PlastiPure, which it claimed were spreading false information about Tritan to generate demand for their own services. Eastman's lawyers asked the judge to bar both firms from ever claiming Tritan was estrogenic or saying that cell-based tests could detect estrogenic activity, even though scientists routinely use them for this purpose. In 2013, Bittner argued his case against Eastman in a federal courthouse in Austin. Eastman's attorneys claimed running a company that tested products for estrogenic activity, as well as one that helped companies find non-estrogenic alternatives, created a conflict of interest. Without directly challenging the validity of Bittner's findings, they leaned on the questionable industry claim that tests based on human cells are not sufficient to establish estrogenic activity. The long legal battle has depleted CertiChem and PlastiPure's coffers.

"We have laid off half of our staff," Usey told me. "It has pretty much crushed us, and emboldened Eastman."

Bans have cropped up around the world, with more than ninety studies examining people with various levels of exposure suggest BPA affects humans much as it does animals. Still, Bittner is not giving up the fight. CertiChem and PlastiPure appealed the Eastman ruling. In 2013, a federal jury sided with Eastman and the court told PlastiPure and CertiChem to change their marketing tactics. In December 2014, an appeals court upheld that ruling. As a result, Bittner's companies have changed their tactics a bit.

"We don't talk about Tritan, or Eastman, in a commercial context concerning the testing results that we have," he says. "But that doesn't limit our discussing our research in a scientific context."

Bittner and his companies are getting their message out by publishing scientific papers about estrogenic plastics that specifically mention Tritan and products made with it. According to Bittner, this tactic seems to be effective.

"I think people are recognizing that it's not the courts that determine scientific questions."

The struggle for consumers' hearts, minds and wallets goes on. Eastman reports that sales of Tritan continue to grow. PlastiPure has announced plans for a new line of baby bottles that meet its own definition of EA-free. PlastiPure continued to receive much positive press coverage during this time but was very much cash-limited to do much more than survive. Many smaller companies were interested in talking with PlastiPure, but they were undercapitalized to bring PlastiPure's safer products to market.

Where's the startup now?

PlastiPure develops environment-friendly plastic materials, processes, and products. It is a technology company focused on developing safe plastic materials and products. It works with innovative companies throughout the plastic supply-chain to develop PlastiPure-Safe EA-Free materials, compounds, colorants, processing aids, and PlastiPure-Certified products. Additionally, PlastiPure educates consumers on the facts. PlastiPure raised $1.1M from the National Institute of Environmental Health and Safety.

Many discussions with potential commercial partners and some co-development work on safer products are ongoing. The main reluctance of commercial clients to invest heavily is concern about liability of their "unsafe" existing product lines. They fear they will be sued for what they have done in the past when they launch safer products. A precedent has been made for this when consumer product companies were sued after they launched BPA-free products. Recently, external investors have been less interested in PlastiPure's licensing model, but in PlastiPure making safer products itself.

PlastiPure decided in late 2014 to launch a safer products' company called productpure to address the dearth in the market of safer consumer products that do not leach hormonal chemicals. PlastiPure created this company in early 2015 and has received much interest from consumers, consumer groups, the popular press, and investors on getting safer products to market. PlastiPure has now completed design of the world's first certified EA-Free baby bottle and the molds are being made. PlastiPure is signing up partners to help pilot this product and has a large network of press and consumer group supporters eager to promote this safer product when it is ready for retail sale.

PlastiPure hopes to soon close an investment round for growth considering the increased cash flow needs of a product company versus the current licensing/consulting company. PlastiPure anticipates that investment will quickly follow a successful and highly visible launch of productpure's first product, an EA-Free baby bottle. PlastiPure will utilize its broad network of supporters to raise awareness that solutions exist (including safer plastics) to the problems that are now being publicized about the dangers of plastics and other materials. PlastiPure perceives its largest risks to be: Insufficient capital to address infrastructure and growth of this new company, productpure, if sales pickup rapidly; and potential future lawsuits by companies that lose market share to productpure.

Case Study Questions

1. Discuss key challenge: build a viable business around a problem consumers' do not know they have, and raise awareness given the company's limited resources
2. Was potential social impact a good reason for NSF to fund this startup?
3. How best to assess target markets – from baby bottles to NGOs
4. Determine strategic fit via partnerships, and discuss partner relations – communications/clarity and expectations alignment
5. Market factors – non-existent (create one) versus existing (carve out niche)
6. Identify areas where the company needs help
7. Discuss regulatory issues, growth and exit strategy
8. Address the issue of entrenched powerful players reluctant to forego large profit margins in public-health arena
9. Take-away(s)

21 Fabrics That Can Remember

Technology Sector: Medical Devices
Startup: MedShape, Inc.
Website: www.medshape.com
Location: Atlanta, Georgia

Federal Funding Timeframe: July 2013 – June 2014
Funding Amount: $0.18M

This narrative is a story of: Private equity funds; FDA approval requirements; IP play; significant societal impact; scale-up/production challenges; unlike all other narratives in this text, the company was not a recipient of additional funding from NSF after the first round

Company and Team

MedShape was founded as a Colorado C Corporation in June 2005. The firm's early origins stem from research performed by Professor Ken Gall then at the University of Colorado at Boulder. From 1999-2005, Dr. Gall secured over $4M in grants to drive fundamental research in shape-memory alloys and polymers. Among other research findings, these efforts culminated in the development of a unique shape-memory polymer specifically formulated for use in permanent orthopedic implants. Ken collaborated with orthopedic surgeon Dr. Reed Bartz and surgical podiatrist Dr. Doug Pacaccio to explore novel surgical applications for shape-memory materials that led to the startup's initial product concepts and working prototypes. To accelerate product development the company was able to recruit a full-time team, including its current President and CEO Dr. Kurt Jacobus, upon successful close of its first private placement round in 2006.

MedShape's vision is to develop distinctive medical devices based on a platform of "smart" materials and technology. It now has several science-driven material technology platforms to create medical devices that utilize the unique capabilities of a new generation of biomaterials to further orthopaedic surgery. These devices created from proprietary shape-memory biomaterials fusing creative design with surgical expertise can adapt inside the body and dramatically improve the technology of soft tissue fixation, fracture repair, and joint fusions. This standalone medical device company, working with strategic partners, has successfully developed market-driven products for distribution. The company manufactures some of its product offerings in an in-house controlled environment using medical manufacturing equipment for custom components, assembly, and packaging operations. For operations that the firm is not currently equipped to manage it seeks out contract manufacturers thus allowing it to focus on its core competencies and reduce overall capital expenditures related to setting up manufacturing processes that it cannot support with the equipment on hand.

The 25-member team has expertise in orthopedic medicine, biomaterials, engineering, regulatory affairs, early-stage company management, medical device development and commercialization, manufacturing scale-up, marketing, and sales. Kurt Jacobus, Chairman, President, and CEO obtained a doctorate in mechanical engineering from the University of Illinois and spent five years as a management consultant with McKinsey & Company. He has over twelve years of early-stage company experience, leading or supporting the start of over a dozen new businesses that today have combined annual revenues of over $50M. He also has a strong background in shape-memory metals.

Jack Griffis, Vice President of R&D, is the recipient of five national design excellence awards in medical device engineering and has been awarded thirty-five patents in biomedical technology with an additional fourteen pending. He is also a recent inductee into the National Academy of Inventors. With over twenty-two years of experience in the medical device industry, Jack has obtained Food and Drug Administration (FDA) approval for more than forty-five Class I, Class II, and Class III medical devices. He has a graduate degree in engineering from Georgia Institute of Technology.

David Safranski has spent nine years researching shape-memory materials and published several studies and book chapters on the structure-property relationships of shape-memory polymers. David received his Ph.D. from Georgia Institute of Technology in materials science and engineering.

Kenneth Dupont is responsible for operations and quality and has ten years of experience in orthopedic bioengineering. He has authored over a dozen peer-reviewed studies in musculoskeletal tissue engineering. He recently obtained FDA approval for his first Class-II orthopedic implant.

The company's directors and advisors include Paul Hills, Chairman of Hills Capital Management. Paul spent thirty-five years as an executive and co-founder of Sage Products, Inc., a medical device marketing and manufacturing company. Since its inception in 1971, Sage Products has developed an international reputation as an innovator of products that provide superior patient care. It was acquired by Madison Dearborn Partners in 2012.

Niles Noblitt is a founder of Biomet, Inc. where he served as the head of engineering and product development before becoming the President of EBI Medical Systems after its acquisition by Biomet in 1988. He oversaw the growth and development of EBI for Biomet before being responsible for a joint venture between Biomet and Merck in 1998. In early 2000, he returned to serve as Chairman of the Board of Directors. Niles retired from the company after its acquisition by the Blackstone Group for $11.4B. Along with his supporting role at MedShape, Niles serves as a board member for NICO, an Indianapolis-based company that has produced a minimally invasive medical device for the removal of tumors and tissues from the human brain.

Jones Day, an internationally recognized law firm, provides general corporate legal counsel including transaction support. Furman IP Law and Strategy supports the company in filing and administration of intellectual property that is central to the firm's success.

Market Opportunity

Funding from the NSF SBIR program is helping MedShape develop a novel medical device to improve surgical outcomes for rotator cuff repair. Over 400,000 such procedures are performed each year in the U.S. at an estimated annual cost of $474M not including hospitalization costs, lost wages, or costs associated with rehabilitation that are significantly larger than the cost of the medical device. Due to tear size, time from injury to repair, the surgical technique used, and tendon-muscle quality, the failure rate of rotator cuff repairs approach 50%. If a tear is left untreated or becomes chronic, the tendon-muscle unit gradually undergoes muscle atrophy and fatty infiltration due to poor biological healing, reduced mechanical properties, high re-tear rates, and poor clinical outcomes. These are irreversible structural changes in the tendon-muscle quality and can lead to irreparable tears in around 30% of cases. Many devices and techniques such as reinforcement patches, complete or partial repair, and tendon transfer, are used as remedies. However, none can reverse muscle atrophy and stop fatty infiltration. Initial clinical studies indicate no consensus on the efficacy of patches since repairs with some biological patches have re-tear rates of 90%. The number of cases is expected to increase due to an aging population's increased desire to lead active lifestyles, and the increasing popularity of "extreme" sports.

Currently, there are no shape-memory fabric-based patches for rotator cuff repair. Biological reinforcement patches have been used for over five years, but these are expensive and have not gained widespread acceptance. While they are biodegradable, an acute inflammatory response often occurs with these devices. Also, mechanical properties cannot be tailored as these materials are mechanically inactive. Synthetic reinforcement patches are less expensive than biological patches, but their mechanical integrity can drastically decrease once exposed to the in vivo (animal model) environment, thus they are only useful for initial suture fixation. The overarching challenge with current patches in rotator cuff repair involves the need to improve tendon-muscle quality by stabilizing the repair site from micro-motion and muscle contractions. Without this the repair is likely to re-tear even with a biological patch. The patch must be mechanically durable and must not elicit an adverse biological response during its implantation. Lastly, the device should be commercially competitive with the current products on the market.

MedShape believes its rotator cuff repair device holds great promise, and that with successful regulatory clearance and market entry the path for introducing new medical devices based on its core materials technology will be dramatically streamlined. Potential follow-on device opportunities include Achilles' tendon repair, muscle atrophy related diseases, cardiovascular repair such as bridging defects and mechanically supporting damaged blood vessels or delivering therapeutic agents, and bridging defects in soft tissues like intestine, skin, and other connective tissues.

The market for rotator cuff reinforcement devices is competitive. Wright Medical and DePuy are the largest players – their market share totals about 80%. Both employ a channel management strategy with broad and relatively commoditized product lines rather than a narrow portfolio of distinctive products. This allows them to own these market channels by ensuring that the independent device distributors who are central to sales in the U.S. and Europe find it difficult to go to market without at least some of their products. This allows the two firms to pursue exclusive relationships with distributors - relationships that limit competition and secure their place in the market. While these strategies have proven successful, the focus on channel instead of technological innovation leaves them vulnerable to devices based on new and disruptive technology. MedShape believes its shape-memory polymer technology will be a distinctive and disruptive product and will be competitively priced. Since the company has already launched two medical devices from SBIR funding, the cost associated with launching a new device will be reduced as the company already has shipping, receiving, quality controls, and distribution channels in place.

Technology and Product

Porous materials are micro-structured to promote adjacent tissue in-growth. The use of porous polymers has been limited in orthopaedic load-bearing applications due to the loss in mechanical properties typically associated with introducing porosity in a material. While porous metals are used for clinical devices, MedShape is the first company to develop and receive FDA clearance for devices manufactured with porous polyether

ether ketone (PEEK) material[27]. Unlike other porous polymer materials used clinically, the company's biomaterial uses a proprietary processing method that seamlessly connects a porous surface to a solid base without compromising the mechanical integrity of the device. Developed by a group of engineers and researchers at Georgia Institute of Technology, devices manufactured from this material are biocompatible, biostable, radiolucent, and magnetic resonance imaging (MRI) safe. Prior to the invention of MedShape's high-strength shape-memory polymer, researchers had not considered such devices for use in load-bearing orthopedic applications.

Shape-memory alloys (SMA) have a history of successful human implantation in biomedical devices such as self-expanding cardiac stents, guide wires and orthopaedic staples. Shape-memory polymers (SMP) are a relatively new class of "smart" materials. They can "remember" multiple shapes and transition easily between those shapes when triggered to do so. SMPs can deform up to 400% and still recover their original shape without loss of mechanical integrity. Appropriate triggers for shape change include heat, light, and mechanical force. This biomedical polymer allows devices to enter the target surgical site in a compact geometry and then be triggered to deploy, with minimal mechanical force, into the optimal geometry for fixation.

The polymer is first synthesized into a permanent shape by standard polymer processing techniques. Subsequently, the polymer is heated above a critical temperature and thermo-mechanically deformed into a temporary shape, a process known as programming. The polymer will remain in the programmed shape until it is heated to a specific temperature, upon which it will experience controlled shape recovery.

Shape-memory polymers have the potential to significantly impact implants and minimally invasive surgery since simple and reliable actuation is often needed in the restricted and highly variable bodily environment. A "smart" suture, for example, has been developed that automatically closes wounds to a tailored force level with minimal intervention.

The goal of the NSF SBIR-funded project is to collectively address the issues in rotator cuff repair patches by creating a novel solution using shape-memory fabric technology. A key requirement to improve the tendon-muscle quality and reduce the occurrence of rotator cuff re-tears is to stabilize the insertion site for proper healing. MedShape's proof-of-concept solution employs a self-deploying shape-memory fabric device that is attached to the rotator cuff insertion site. It changes shape slowly at body temperature with the capacity to apply a counterforce to the tendon-muscle unit, thereby stabilizing the repair site. The device would be delivered in a stretched form and then deployed slowly back to its original contracted shape via body temperature and fluid uptake over days to weeks depending on the application.

Once deployed, the device's gradual and continual contraction would provide tension against the tendon-muscle unit, which reduces the amount of motion at the repair site and may prevent muscle atrophy and fatty infiltration. Fabric properties and structure would be tailored to prevent suture pull-through, to contract at a varying-force rate to avoid tendon tear, and to be equivalent to the initial mechanical properties of other synthetic patches. The surgical procedure for attaching the SMP fabric would follow established arthroscopic protocol for inserting current rotator cuff patches. This approach has several advantages compared to current rotator cuff patches such as being mechanically active with tailored mechanical properties, having a microfiber structure that is being easy to fabricate and with the potential to be biocompatible.

In 2015, MedShape intends to finalize device design and perform manufacturing transfer. With a compliant design and the development of manufacturing processes, clearance for human use is secured through the submission of an FDA application that will include detailed technical device descriptions, laboratory, and in vivo testing of performance relative to other cleared products, biocompatibility, long-term storage and stability trials, and an exhaustive literature review.

On average, the regulatory clearance process for Class II devices takes 2-3 years to complete and will cost $3-6M per product. This large investment of time and energy establishes a high barrier to market entry, and often requires that new medical device products have a strong IP position to protect them and allow long-term profits. Once the FDA 510k clearance is granted, MedShape will pursue further uses and combination devices based upon this shape-memory fabric platform, such as therapeutic agent release from the shape-memory fabric itself.

Since its earliest days, the firm has endeavored to construct a powerful IP portfolio around its devices that use shape-memory fabrics. Its IP strategy is to leverage both patents as well as trade secrets, with the latter being reserved for valued inventions for which determination of infringement presents challenges. This strategy also seeks to establish IP across the entire production-and-use cycle for its products, thereby erecting multiple barriers to entry for competitors, by employing both broad claims addressing multiple markets and narrow claims focused on specific designs and targeted market segments.

MedShape has nineteen issued patents and eight pending patents on the design and implementation of shape-memory materials for orthopedic applications. The company has broad claims to the use of shape-memory fabric in orthopedic soft tissue reinforcement and elongation, including thermo-mechanical properties and fabric structure. It also has many trade secrets relating to the synthesis, tailoring, processing, manufacturing, deployment, and packaging of shape-memory polymers for biomedical devices.

Business Model and Execution

The business model is to develop distinctive orthopedic medical devices based on MedShape's platform of shape-memory materials technology. Currently, the company is developing various implantable materials used in orthopedic applications such as soft-tissue fixation, trauma, and fusion. It has successfully developed, cleared, and is selling orthopedic implants. In the next 1-3 years, the firm is focused on rapid growth in sales of its five FDA-cleared products in both the U.S. and international markets. It currently utilizes a combination of direct sales and distributors as sales channels. Two of the five medical devices were supported by NSF and NIH SBIR funding and marked the first FDA-approved shape-memory polymer medical devices.

MedShape plans to expand in almost all areas of biomaterials. This includes the support of ongoing sales of its current products, expansion of its sales and marketing capabilities, FDA clearance of additional products based upon its core technologies, and the development of strategic partnerships. It will promote its "smart" materials platform to secure licensing revenue and non-dilutive funding opportunities to pursue further growth beyond the current orthopedic applications being pursued. Over the subsequent 4-8 years, MedShape will identify, develop, obtain FDA approval, market, and sell new medical materials and medical products based on shape-memory technologies. It plans to leverage its experience in developing novel shape-memory devices into other arenas of "smart" materials such as bio-functional, biodegradable, and therapeutic materials.

Since its founding, MedShape has completed five rounds of equity financing to support its growth. In 2006, it closed seed financing of $1M to further develop its initial slate of products focused on orthopedic implants. This was followed by a $3M investment in 2008 to finalize development and FDA approval of the company's first polymer-based product for shoulder rotator cuff repair. In 2009, it closed $10M in equity financing to finalize FDA clearance of three products as well as expand its biomaterials research capability to new applications. The fourth round of $6M in equity financing was to complete the launch of two more products and to further fundamental materials research into biopolymers. The most recent round consisted of $11M in new financing was to complete the launch of its *Eclipse* product, to soft launch the new *ExoShape Femoral* device, and to further develop and commercialize other products in the pipeline. To date, the company has raised $30M in equity financing, garnered $3.5M in licensing revenues and obtained grants worth $3.9M. Total NSF SBIR funding to the company in the period 2007-2014 has been approximately $1.5M. Federal grant funding resulted in two commercial products. The successful commercialization of these technologies culminated in the company receiving the Small Business Administration's 2011 Tibbett's Award.

The launch of MedShape's first two products was successfully completed in 2010. To support the commercialization of these devices, the company completed a facilities expansion to accommodate operations, such as shipping and receiving, as well as building a sales force to cement key relationships with clinicians. The firm also received FDA clearance to launch its *ExoShape* soft tissue fastener, a shape-memory polymer implant used primarily for tendon and ligament reconstruction for knees and shoulders. More recently, the company received FDA clearance to launch its *DynaNail* product, a shape-memory metal alloy implant used to fuse damaged or diseased joints and bone fractures. In 2013, the firm received FDA clearance to launch its *Eclipse* soft tissue fastener, another shape-memory polymer implant designed to accommodate surgery in the hands, elbows, and feet. In 2014, MedShape received FDA clearance to launch its *ExoShape Femoral* soft tissue fastener. In the period 2010-13 the firm has increased its annual sales revenue to approximately $5M proving it has the capability to ramp sales and provide clinical support for technically complex medical products. Although it has begun to see successful market penetration for its other Class II medical products, sales revenue is not yet sufficient to support operations and large-scale preclinical animal trials.

The company believes its most likely exit will be an acquisition by a strategic partner in the medical device space. Acquisition can take many forms and may be limited to specific clinical specialties such as sports medicine, or for example a suite of products specific to the foot and ankle.

NSF Commercialization Assistance and Impact

MedShape was introduced to IA in November 2009 by their NSF Program Director. The company's needs as identified by IA converged upon the company's commercialization efforts. Contemplated issues include introductions to financial companies and investment bankers, scouting the landscape of opportunities, and to be a sounding board for financial negotiations and terms. IA activities and outcomes are summarized in the following Table.

IA Activity	Outcomes
Startup Overall Engagement	18 months; 20+ hours of interaction; 75+ e-mails sent to and on behalf of MedShape
Operational Assistance	Offered help to CEO in various operational areas; provided financing introductions and guidance
Introduce Potential Partners	Biomet
Find Investors	Capstone Advisory Group; Commonwealth VC; In-Q-Tel
Identify Domain Experts	Michael Zarriello; Kevin Dewalt

Where was this startup at the end of federal government funding?

MedShape is an example case study where IA aided the company but that did not result in the company being awarded a Phase II grant from NSF. The panel of experts, both technical and commercial, that convened at NSF to consider MedShape's proposal identified several weaknesses as outlined in the Table below.

• Not clear that the desired shape-memory effect can be achieved; without the shape-memory effect, this cannot be considered an innovative technology
• Not clear that sufficient resources will be allocated to optimize the shape-memory effect; it appears that the proposed objectives are to fulfill requirements outlined in a FDA guidance document for surgical mesh devices and to support a future regulatory submission
• The need for follow-up surgery to remove fabric implant is a significant drawback
• It is not clear whether cyclic loading from joint movement will adversely affect implant function
• Weaknesses of the proposed animal studies are lack of control group using commercially available surgical mesh; 3-month follow-up likely not sufficient to observe whether stable repair has been achieved
• Licensing terms with Colorado State University unclear
• Based on the amount of capital raised, the ownership status is unclear

Where's the startup now?

MedShape develops science-driven medical devices that utilize the unique capabilities of a new generation of biomaterials. MedShape has several innovative material technology platforms that convey its vision, expertise, and passion to continue to evolve orthopedic surgery. Shape memory polymers (SMP) are finding their way into everyday life. Potential new applications of SMPs in the automotive industry include self-repairing fenders that use heat from a household hairdryer to remove dents. Appropriate triggers for shape change include heat, light, and mechanical force. Shape memory polymers can "remember" multiple shapes and transition easily between those shapes when triggered to do so. SMPs can deform up to 400% and still recover their original shape without a loss of mechanical integrity.

MedShape, Inc. is a privately held medical device company working to develop and commercialize a portfolio of surgical solutions that use its patented shape memory polymer and metal alloy technologies to address the increasing demand for improved sports medicine, joint fusion, and musculoskeletal trauma products. It develops and markets soft tissue fixation and fracture management devices. It develops shape memory medical devices, including spinal cages and suture anchors for orthopedics. The company was founded in 2007 and is based in Atlanta, Georgia. As of 2013, the company has raised $36M in funding, the latest investment round involved $11M in additional financing. MedShape, Inc. is the industry leader in shape memory orthopedic devices. The investment will help drive further development and commercialization of pipeline products, including the full market release of the *Eclipse Soft Tissue Anchor*. MedShape recently announced its newest shape memory fixation device for soft tissue repair.

Kurt Jacobus, CEO of MedShape - "This latest round of financing represents another significant milestone as MedShape continues its growth in the joint fusion and soft tissue repair segments. The launch of *Eclipse* indicates the company's continued dedication towards advancing orthopedics by developing innovative devices using shape memory technology."

Made from MedShape's proprietary *PEEK Altera*, *Eclipse* features a non-rotational deployment technique that allows surgeons to better replicate native soft tissue anatomy and functionality. The two-part sheath-and-bullet design provides a simplified insertion and improved fixation strength compared to traditional tenodesis screws. Available in different diameters and lengths, *Eclipse* accommodates numerous anatomies in tenodesis and tendon transfer procedures. MedShape received FDA 510(k) clearance for *Eclipse* in March 2013.

Since August 2013, *Eclipse* has been under a controlled market release, available to a select group of surgeons. The first clinical use occurred on August 5, 2013, with Dr. Grant Padley successfully performing a proximal biceps repair at Arizona Spine and Joint Hospital in Mesa, Arizona. In addition, *Eclipse* has been successfully used in hand, knee, foot, and ankle soft tissue reconstruction procedures.

"I prefer the *Eclipse* device over the implant I previously used for its ease of use," said Dr. Padley. "Unlike traditional tenodesis devices that can rotate and lacerate the soft tissue, *Eclipse* sheath expansion method gives me the ability to place the soft tissue in the prepared bone tunnel and easily maintain the desired position throughout the procedure."

Additive manufacturing (3-D printing) is an emerging materials processing technology. This novel technology removes any design constraints due to the manufacturing process and provides complete flexibility to design a medical device specific to its desired function and application. 3-D printing allows for devices to be fabricated with complex geometric features (e.g., internal channels, porosity) or customized to match a patient's specific anatomy. 3-D printing can also improve manufacturing efficiency enabling devices to be fabricated in small batches at a lower cost with quicker turnaround. Despite 3-D printing's numerous advantages, few 3-D printed metal orthopedic devices have reached clinical use due to the negative impact the fabrication process imposes on the material's mechanical properties, putting 3-D printed devices at risk of failure in many load-bearing orthopedic applications. MedShape is one of the first companies to develop and FDA clear a titanium alloy bone plate utilizing a novel 3-D printing process that maintains the mechanical properties close to the original bulk material.

Nickel-titanium (NiTi, Nitinol) is the most used SMA and can recover strains up to ten times more than traditional metals and alloys. SMAs can change their shape up to 8% and still fully recover their original geometry. Fixation devices incorporating shape memory alloys can respond to local changes in the site of implantation, such as bone resorption, maintaining apposition of bony fragments and sustaining compression across fractures or fusion zones. MedShape is the first company to obtain FDA clearance for a bone fusion device comprised of both titanium and nickel-titanium, paving the way for a range of devices that are both strong and dynamic.

Case Study Questions

1. Comment on NSF SBIR program decision not to fund MedShape's Phase II proposal
2. Could IA have done different in aiding that resulted in a positive outcome to fund?
3. What are the specific challenges for a U.S.-based medical device startup company?
4. Identify risk elements (technical, team, market, finance) and how to manage/mitigate them?
5. How did the company fund this product development without further NSF SBIR funding?
6. Comment on company's growth and exit strategy
7. Discuss the mix of equity financing, licensing revenues, and grant funding
8. Take-away(s)

22   Can You Believe I am Walking!

Technology Sector:	Robotics
Startup:	Ekso Bionics, Inc.
Website:	www.eksobionics.com
Location:	Richmond, California
Federal Funding Timeframe:	January 2007 – September 2013
Funding Amount:	$1.37M
This narrative is a story of:	An IPO; private equity funds; joint development agreements; multiple grants; significant societal impact; scale-up/production challenges

Company and Team

Berkeley ExoWorks was founded in 2005 by Homayoon Kazerooni, Russ Angold and Nathan Harding, all members of the Berkeley Robotics and Human Engineering Laboratory at the University of California, Berkeley (UCB), to commercialize groundbreaking technology in the field of human exoskeletons. This laboratory has a long history of original design and control approaches for robotic systems that interface with humans. Engineering personnel, many of whom UCB graduates, have strong experience in electromechanical design, control, real-time programming, and human-machine systems. Human exoskeleton technology is in its infancy and most research in this area is currently conducted in Japan and America. The company's mission is to develop and manufacture powered exoskeleton bionic devices that can be strapped on as wearable robots to enhance the strength, mobility, and endurance of paraplegics and soldiers.

In 2007, the company changed its name to Berkeley Bionics while developing the *Human Universal Load Carrier (HULC)*, an untethered, hydraulically powered exoskeleton designed to carry heavier loads than previous models. This product was publicly announced in 2009 when an exclusive licensing agreement was signed with Lockheed Martin for further military development. The company also introduced its first commercial product, the *Exoskeleton Lower Extremity Gait System (eLEGS)*, an intelligent exoskeleton that enables wheelchair users with any amount of lower extremity weakness, including paralysis, to stand up and walk. In 2011, the company changed its name yet again to Ekso Bionics and rebranded the eLEGS product as *Ekso*. In 2012, Ekso received approval from the FDA for hospital use in the U.S. and received a Conformite Europeene (CE) Marking from the European Union.

Ekso was selected as Wired magazine's "Second Most Significant Gadget of 2010", was included in Time magazine's "50 Best Innovations of 2010" and featured in Inc. magazine as one of "5 Big Ideas for the Next 15 Years". The company currently sells Ekso designed to provide upright gravity-dependent exercise to patients. It has expanded its executive team to address the needs of an eighty-person, fast-growth company, and to help meet regulatory requirements for medical devices sold in the U.S. and European markets. The team consists of:

- Eythor Bendor, CEO, has over fifteen years of experience in the medical device industry. He previously was the President of Ossur Americas and before that the CEO of REX Bionics.
- Max Scheder-Bieschin, CFO, was the cofounder-CEO of Barefoot Motors and has seventeen years of investment banking experience including time at Deutsche Bank.
- Nathan Harding, COO, has extensive experience creating and managing teams in environments where multiple technologies intermingle. At Berkeley Process Control, Nathan led a team responsible for creating automated equipment products, which grew from $0 to $30M/year in sales.
- Russ Angold, CTO, has a diverse engineering background developing patented products in numerous industries. Prior to Ekso Bionics, Russ developed new irrigation products for Rain Bird, and before that worked at Berkeley Process Control designing and developing high-speed optical fiber manufacturing machines and wafer handling equipment.
- John Fogelin, Vice-President (VP) of Engineering, has twenty-five years of real-time computing experience including experience as CTO and VP of Engineering at Wind River Systems.
- Kolbeinn Bjornsson, VP of Sales, was CEO of HRV Engineering and served as VP of International Sales for seven years at Ossur.
- Karl Gudmundsson, VP of Marketing, has twelve years of product management experience in the medical device market at Biomet and Ossur.
- Frank Moreman, VP of Manufacturing, has ten years of executive experience in the medical device and semiconductor manufacturing industries. He was also the owner and COO of Sieger Engineering.
- John Tugwell, Director of Quality Assurance and Regulatory Affairs, has twenty-five years of experience obtaining FDA and CE certifications for Class I, II, and III medical devices.
- Andy Hayes, Sales Director for Europe, Middle East and Africa at Ossur, has over twelve years of experience in medical device sales and support.
- Renee Loth Cali, Director of Culture, has twenty-one years of

human resource expertise including recent experience at startup medical device company, Intuitive Surgical.

The Board of Directors:

- Scott Banister, an early investor, and Director at PayPal is the Chairman and cofounder of IronPort. Scott was the lead investor in Ekso Bionics' Series A-1 financing round.
- Jack Peurach, Executive VP of R&D at SunPower, a medium-size solar-technology company in Richmond, California. Jack represents the interests of the founders on the Board.
- Michael Fawkes was the Senior VP of Global Operations at HP.
- Marilyn Hamilton was cofounder of Quickie Wheelchair later sold to Sunrise Medical.
- Eythor Bendor, CEO of Ekso Bionics.

Market Opportunity

In the U.S., 5.6M people currently have impaired mobility from several different causes. Ekso Bionics is initially focusing on the spinal cord injury (SCI) portion of this market largely because this population is made up of younger patients with more consistent limitations in general than patients with limited mobility from other causes. The current U.S. SCI population is nearly 300,000 with an estimated 12,000 new traumatic incidences each year. Similar population sizes are found in other countries around the world as well. In addition, there are 26,500 U.S. rehabilitation facilities that include in-patient centers (5%), out-patient centers (60%), and post-rehabilitation care centers (35%).

SCI patients experience two primary market forces - prohibitive costs and reduced quality-of-life associated with extended care. It is estimated that an SCI patient is liable for around $1.5M in direct costs and $2.5M in indirect costs over their lifetime. Much of these costs are not associated with the primary injury but are instead related to the continuing care of common secondary health risks associated with the physical deconditioning caused by limited mobility. The secondary effects of limited mobility include muscle atrophy, impaired skeletal and circulatory health, diabetes, and obesity. The leading cause of death is renal failure typically attributed to the large amount of medication patients take to combat these secondary health risks.

Beyond direct physical impacts, the loss of gravity-dependent ambulation compromises quality-of-life, hinders performance of everyday activities, and has a greater negative impact on overall health than nearly all other issues associated with paralysis. Research indicates that maintaining lower extremity physical condition is an effective countermeasure to reduce the impact of these secondary health risks. Thus, a device that can provide SCI patients with access to extended duration gravity-dependent exercise to directly combat the risks associated with physical deconditioning could help make significant improvements in the cost of care and quality-of-life. The existing Ekso device seeks to meet this primary need - the earlier device was however only capable of providing this exercise under medical supervision.

Approximately 500,000 Americans survive a stroke each year. The primary means of mobility for such patients is the wheelchair as it has been for the last fifty years. Despite the benefits introduced by widespread use of wheelchairs, a significant need exists for a device that can provide standing mobility inside a patient's home to maneuver through tight rooms, ascend stairs, and even stand at an unmodified stove to prepare meals. Such a device does not need to replace all the functions of a wheelchair, in fact, the first generation of mobility exoskeletons will likely be used in tandem with a wheelchair to meet patient's mobility needs and expose them to gravity-dependent exercise. Miraculously, most stroke survivors can effectively relearn skills like walking if they are aided in making the correct motions by a machine or a physical therapist while part of their body weight is supported. This training is expensive - it requires the patient to make regular visits to a qualified physical therapy center.

Ekso Bionics is developing an in-home training device that allows a post-stroke patient to undergo rehabilitation with little or no professional assistance. It will offer patients the ability to progress from walking directly after injury to walking out the door. Patients can relearn ambulation in the privacy of their homes with help from family members or friends. Though many organizations work on different treadmill devices for post-stroke gait training, there are currently only two competing devices in the market - *Rewalk* from Argo Medical Technologies and *REX* from REX Bionics. Neither is designed to facilitate gait rehabilitation and as a result have yet to be adopted by the rehab community. The powered orthotics market is new and emerging. Thus, while the company seems to have a strong market position, success will be determined by the ability of its devices to expand their influence, technology, and reach as the market continues to mature.

The rehab market requires more sophisticated control to work with partially paralyzed limbs as the patient develops more and more of their own skill. Ekso Bionics believes that its expertise in control will differentiate it from the competition. Additionally, the company's expertise in exoskeleton design can produce a much smaller human exoskeleton that will offer patients the ability to easily enter and exit vehicles and tackle other facets of everyday life. The product is lightweight, unobtrusive, and can be taken home to train patients while they go about their daily lives. The product will further differentiate itself by allowing the patient to sit down into typical seats such as those found in automobiles, restaurants, and trains. The device is offered in two versions: one for those with an affected left side and one for those with an affected right side.

Technology and Product

Designs for two products, the *ExoHiker* and the *ExoClimber*, each with a weight-carrying capacity of one hundred and fifty pounds, were completed in 2005. The first was intended to help hikers carry heavy loads on their back. The almost-silent, battery-powered skeletal system weighing thirty-one pounds can be strapped on to a hiker's body and controlled with a handheld LCD display. It can operate at an average speed of 2.5 mph for forty-two miles using an 80-watt-hour lithium polymer battery weighing 1.2 pounds. With a small solar panel, its "mission time" can be unlimited. The second product allows the wearer to ascend stairs and climb steep slopes. It weighs fifty pounds, and for each pound of lithium polymer battery, can assist a climber to vertically ascend six hundred feet with a 150-pound load. The previously mentioned *HULC* can carry a 200-pound load and reduce metabolic energy needed by the wearer to perform a given task. This product lends itself to augmentation with devices that can be mounted on the back of its exoskeleton. One such device, the *Lift Assist Device*, lets operators carry front loads as well as loads on their back. It also allows single operators to lift heavy loads that currently require two or more people.

Ekso, another hydraulically powered exoskeleton system, allows paraplegics to stand and walk with crutches or a walker[28]. It is an in-home gait training device, a human exoskeleton that allows post-stroke patients to undergo rehabilitation with little or no assistance. It will be a cost-effective system that can be leased from a physical therapy center and taken home to provide higher quality gait training to stroke victims who do not live near a stroke rehabilitation center or who need more training sessions than can currently be offered through office visits. Its computer interface uses force and motion sensors to monitor the patient's gestures and motion to help interpret the intent of the user and translate it into action. Patients can by themselves put on and take off the device as well as walk, turn, sit down, and stand up unaided. It weighs forty-five pounds, has a maximum speed of 2 mph, and a battery life of six hours. It can help transfer a user from a wheelchair to a chair and is suitable for people weighing up to 220 pounds, who are between 5'2" and 6'4" tall. It allows the user to walk in a straight line, stand from a sitting position, stand for an extended period, and sit down from a standing position. This product is undergoing development and clinical trials in rehabilitation centers.

Utilizing recent breakthroughs in human exoskeleton design, *Ekso* is a lightweight, portable exoskeleton that cradles a patient's lower extremities and torso and maneuvers their paralyzed limbs for them. Locomotor training helps stroke survivors relearn skills that are lost when part of the brain is damaged. Typically, during training, one or two physical therapists help survivors regain the use of stroke-impaired limbs – this requires focused repetitive practice. New, automated rehabilitation systems have recently been developed that reduce the high labor cost of manual locomotor training. These new devices are large and expensive, and patients need to use them in a rehabilitation center. While these rehabilitation systems are confined to these specialized centers, *Ekso* will allow the patient to walk, maneuver and have a longer, more enjoyable rehab experience. Without practical solutions for maneuvers such as donning, doffing, sitting down, standing up, and turning, an exoskeleton gait trainer would not be suitable for in-home rehab use. Several design and control tasks in *Ekso* allow the above maneuvers.

This product design includes a compact, lightweight, and quiet actuation system that has a single integrated motor-pump system able to power both a knee and a hip actuator. It also regenerates power when the exoskeleton leg is resisting motion. Another important feature is a device controller to decide what the exoskeleton trainer should do based on various sensor measurements from the patient-device interface. Characteristics of controllable bodyweight support could move most locomotor training out of the clinical setting and dramatically reduce labor costs. Existing exoskeleton prototypes can be produced for less than most locomotor training equipment reducing capital costs as well. Patients would be able to wear such devices for most of the day, thus remaining mobile and gaining the therapeutic effects of bodyweight-compensated physical therapy over the course of a day, rather than just a short session. Ultimately, creating such a device will give clinicians an alternative to wheelchairs for patients who do not need locomotor training, and provide significant benefits from functioning upright with a sizable load on their bone and/or muscle structure.

Ekso Bionics has a licensing agreement covering human exoskeleton technologies developed at University of California, Berkeley. A strong IP position in lower extremity exoskeletons with several patents in the areas of mechanical architecture, design, control, and human-machine interfaces for industrial and medical applications provides a competitive advantage. The company currently has exclusive control of eighteen patents that are in various stages of prosecution.

Business Model and Execution

Early in the product lifecycle, the company's strategy is to manufacture the entire exoskeleton, and then as the product matures, to serve as an integrator while continuing to manufacture core components. It plans to market the devices to individuals in an in-home environment. The path-to-market uses inpatient rehabilitation centers as a method for interfacing with patients. These centers then serve as an intermediary point of contact, assisting in sizing, prescribing, and training users in exoskeleton use. The devices will require training to certify user safety before they can go home with it. This structure also provides a certified distribution and service network to support new devices in the field. A strong connection to the rehab market facilitates transfer into the home market. In the meantime, rehab centers themselves serve as customers. This partnership arrangement decreases time-to-market and allows rapid product iteration. This strategy allows for a parallel effort for FDA certification for home use and eligibility for Medicare reimbursement.

Partnering with the rehabilitation community is essential on two fronts - clinical practice and research. The company has formed partnerships with leading hospitals to help introduce the technology, incorporate it into their clinical practice, and provide valuable feedback about device-use expectations in standard clinical environments. This requires time spent at each facility conducting patient testing, gathering data, and finalizing regulatory certifications for FDA approval. Product launch was achieved in 2012. The strategy for the home device will build on the success of the clinical device. By providing a contiguous experience that transitions from the clinical to the home setting, the patient gains vital tools for device-use while in the clinical environment. Clinics will train patients on how to properly use the device and customize the hardware for individual patients. This allows evaluation of the efficacy of training procedures and the way patients want to use the device.

The company has successfully implemented a FDA-certified quality control system. *Ekso* has been classified as a Class I medical device by the FDA, but this limits the use of the current device only under medical supervision. The home device will require certification as a Class II medical device. This is of strategic importance as it allows the company to define the classification standard for lower extremity exoskeleton devices that all other future embodiments must meet. Another hurdle involves patients purchasing the exoskeleton being reimbursed by Medicare. This requires that the exoskeleton price be comparable to existing powered wheelchairs. Establishment of Medicare reimbursement codes for *Ekso* purchase is directly linked to demonstrating through clinical evaluation the improved care and reduced costs associated with its use. The company's technology will provide a price/cost advantage over current products since it does not rely on a centralized rehabilitation facility but rather on the distributive use of leased exoskeletons.

In 2012, Ekso Bionics closed a $12M A-2 round of funding led by a $6M equity investment by Chickasaw Nation Industries (CNI). The goal of placing exoskeleton devices in homes was a significant factor in CNI's investment decision. The initial intent was to raise $6M but due to the significant interest generated, the round was kept open to accept more funding from additional participants. While CNI has only an observer seat on the Board, it has helped guide company strategy in terms of moving this technology into the home market as quickly as possible because of the opportunity for greater sales volumes. In 2013-2014 the company was aggressively approached by investment banks to consider an Initial Public Offering (IPO). Ekso Bionics is now listed on the Over-the-Counter Bulletin Board (OTCBB) as Ekso Bionics Holdings, Inc. (EKSO). OTCBB is a U.S. quotation medium used for many over-the-counter equity securities that are not listed on the NASDAQ or a national stock exchange.

Currently EKSO has a market cap of approximately $100M. The company has more than one hundred and fifty international patents with ten U.S. patents granted and another eight filed. This gives it future annuity-like revenue from partners who license the technology. The company announced a partnership with Ottobock, a global leader in prosthetics and neurorehabilitation with subsidiaries in 49 countries and 2013 sales of over $1B. Under the terms of the agreement, EKSO would license two prosthetics technologies to Ottobock in return for licensing and royalty payments. According to EKSO's 2013 annual filing with the Securities and Exchange Commission (SEC), Lockheed Martin paid the company more than $6M in licensing fees for *HULC* alone. Revenue comes not just from licensing but also from product sales, in particular its new *GT* robotic exoskeletons, a total of forty-six of which have been shipped to date. SoldierSocks, a military contractor, has ordered eighty *GTs* over the next three years after its initial ten-suit purchase. EKSO has expanded its global presence shipping a total of 95 units to 72 centers in 17 countries.

In 2013, EKSO was awarded a $1M contract by the U.S. Army to build and test a next-generation exoskeleton. The *Tactical Assault Light Operator Suit (TALOS)* has already been dubbed the "Ironman suit", considered to be at the vanguard of U.S. military innovation. In 2014, it was announced that work in this first phase of the *TALOS* project had won the military's trust to the point where they had decided to award the second phase of the development contract to EKSO. This contract award marks more than $35M in developmental work done by EKSO's team of engineers. The second phase involves the development of an exoskeleton that is both light and strong along with a full range of motion.

In 2015, EKSO announced a study being conducted with nine major hospitals in Europe that will examine how its *GT* suits might alleviate complications of surgeries and even seemingly unrelated ailments like bladder dysfunction that are commonly associated with spinal cord injuries. Early results from the study are expected in 2016 and could help establish the firm's contributions to recovery therapy. In the meantime, its medical device revenue has grown 100% in 2013-14. The company's sales channels are growing deeper, broader, and more profitable and it is winning business away from older, more mature players with dramatically more resources.

NSF Commercialization Assistance and Impact

The NSF SBIR Program Director recommended Ekso Bionics to IA in 2011. Following this introduction, IA worked with the company to understand its short- and long-term goals. They identified the startup's need for investment capital, for developing a fundraising and financial strategy, and for gaining the interest and support of insurers. Accordingly, IA worked with the company to identify key financial, healthcare, and insurance partners; to secure advice and feedback from multiple domain experts; and to introduce its technology to potential customers and strategic partners. IA's introductions led to the successful first round of investment. Additionally, introductions to a handful of world-class investment banks, investment funds, and a potential strategic partner have provided invaluable feedback for future strategies and indications of current market trends. Ongoing conversations with these organizations will assist with capital needs, insurance reimbursement, and other issues critical to success.

In 2011, IA featured Ekso Bionics as a case study at Carnegie Mellon University (CMU) during the "IA @ CMU" event. Additionally, IA along with CMU and the Robotics Technology Consortium (RTC) collaborated to hold a "RoboBowl" Competition. The RoboBowl is envisioned as a series of next-generation robotics venture competitions intended to find and mentor startup and early-stage companies seeking to develop robotic products and services. In doing so, the RoboBowl expects to help address the nation's need to create new jobs and viable businesses by catalyzing the adaptation and commercialization of emerging next-generation robotics technologies.

The competition narrowed the field using a multi-tiered approach with the finalists pitching before a panel of judges invited by IA and chaired by Helen Griener, a founder of iRobot. The RoboBowl competition attracted widespread interest and was attended by, among others, Chuck Thorpe, Assistant Director for Advanced Manufacturing and Robotics, White House Office of Science and Technology Policy, Rory Cooper, Quality-of-Life NSF Engineering Research Center, Venetia Kontogouris, Senior Managing Director of Trident Capital, Richard Lunak, President & CEO, Innovation Works, and Frank DiMeo, VP of InQTel's Physical and Biological Technologies Practice.

IA and Ekso Bionics continued to execute on the plan to secure investment capital from the private sector; assist in securing strategic partners in the rehabilitative care and insurance industries; increase the company's presence in multiple markets segments; and assist with executing the company's long-term commercial and financial strategies. In 2012, the National Innovation Fund, IA's venture partner arm, invested $1M in the company, and assisted the firm in its listing on the OTCBB.

Testimonial

Nathan Harding, co-founder and CEO, Ekso Bionics - "IA's value proposition is worth much more than a dollar amount; expanding opportunities, introducing our team to the right potential investors, business partners and connecting us with industry has been critical not only to our commercial success but to our way of thinking."

Where was this startup at the end of federal government funding?

Ekso Bionics continues to pursue stroke rehabilitation as a new market that complements the original use of the *Ekso* device for spinal cord injury patients. At the outset the goal was to facilitate in-home rehabilitation for stroke survivors, but research during the project suggested that commercialization of a home-use device is not yet feasible. Nevertheless, clinical testing – both at outside hospitals such as that at University of California, San Francisco (UCSF) and internal testing with the company's in-house clinical team – suggests that exoskeletons can hold great potential for aiding in the rehabilitation of stroke patients. As a result, the developments from this project are forming the core of the company's next production exoskeleton and Ekso Bionics expects stroke rehabilitation exoskeletons to be a core business.

Ekso Bionics has already released an exoskeleton (and two different feature upgrades) which is being used by rehabilitation centers to help SCI and stroke patients stand up and walk. This original exoskeleton, *Ekso*, was first sold in 2012 and its development was partially funded by an NSF SBIR project. Two patents resulted from this work; at present both are filed as full utility applications under the Patent Cooperation Treaty (PCT) in preparation for domestic and international filing, and both focus on the control systems that drive the *Ekso*. The first is PCT/US2011/052151, titled "Human Machine Interface for Human Exoskeleton." The second is PCT/US2011/055126, titled "Human Machine Interfaces for Lower Extremity Orthotics."

The company's new *Ekso* device will be named *Ekso GT* and will appear on the market later this year. A provisional patent, Serial No. 61/735,816 titled "Reconfigurable Exoskeleton" has been filed. The patent application covers the mechanical adaptations of the exoskeleton that are related to adapting a device between spinal cord injury and stroke survivors. The company had ten employees at the beginning of this federal funding, and forty employees at the end of this funding, with 2012 revenues totaling $2.6M. The target is rehabilitation facilities where physical therapists use the device to aid in rehabilitation of stroke patients. It is estimated that of 1,200 inpatient rehabilitation facilities (IRFs) and 15,000 skilled nursing facilities (SNF) in the United

States, between 3,000 to 5,000 such facilities are appropriate for the use of *Ekso GT*. While the current device has been well received, clinical partners have expressed interest in the new Ekso GT system which is targeted towards stroke rehabilitation. It is important to have local clinical support because company personnel can repeatedly visit these centers to evaluate changes made in the device. Connections with leading research hospitals, such as the Rehabilitation Institute of Chicago, which is conducting research with the current *Ekso* device and would like to conduct studies within the stroke population using *Ekso GT*, are being maintained.

Where's the startup now?

Ekso Bionics develops and manufactures powered exoskeleton bionic devices that can amplify a person's natural ability and improves their quality of life. This device can be strapped on as a wearable robot to enhance the strength, mobility, and endurance of soldiers and paraplegics The *Ekso GT* wearable exoskeleton is used in over hundred and seventy rehabilitation centers worldwide to help patients with spinal cord injury and stroke, stand and relearn to walk. In September 2017, Ekso Bionics raised $34M in a fully financed rights' offering that included a $20.5M investment by Puissance Capital.

Ekso Bionics Holdings, Inc. (EKSO), an industry leader in exoskeleton technology for medical and industrial use, announced on October 10, 2017 that the company will showcase the latest features of the *Ekso GT* wearable exoskeleton at the 2017 American Academy of Physical Medicine and Rehabilitation (AAPM&R) Annual Assembly taking place October 12-15, 2017, in Denver, Colorado. Live demonstrations of the *Ekso GT*, the first and only exoskeleton cleared by the FDA for stroke and spinal cord rehabilitation, will take place. Clinicians will be able to explore the technology firsthand, learn about the clinical benefits, and engage with company experts on exoskeleton rehabilitation.

The *Ekso GT* enables individuals to stand up and walk with a full weight bearing, reciprocal gait. The device amplifies patient's natural abilities and is designed to help them get back on their feet earlier by aiding their re-learning of correct step patterns and weight shifting. The *SmartAssist* technology is the next generation gait therapy software that allows physical therapy to vary the support of the device for each leg independently – from full power to free walking – and thereby offers personalized therapy specific to the needs of a larger range of patients. *EksoPulse* is an advanced cloud-based technology for physical therapists to obtain insights in rehabilitation to capture patient progress and allows further personalization of care in real-time. The *Ekso GT* with *SmartAssist* software is the only exoskeleton available to rehabilitation institutions that can provide adaptive amounts of power to either side of a patient's body, challenging the patient as they progress through their continuum of care. The suit's patented technology provides the ability to mobilize patients earlier, more frequently, and with a greater number of high intensity steps. To date, this device has helped patients take more than eighty million steps in over 185 rehabilitation institutions around the world.

Ekso Bionics is a leading developer of exoskeleton solutions that amplify human potential by supporting or enhancing strength, endurance, and mobility across medical, industrial and defense applications. Founded in 2005, the company continues to build upon its unparalleled expertise to design some of the most cutting-edge, innovative wearable robots available on the market. Ekso Bionics is the only exoskeleton company to offer technologies that range from helping those with paralysis to stand up and walk, to enhancing human capabilities on job sites across the globe, to providing research for the advancement of R&D projects intended to benefit U.S. defense capabilities. The company is headquartered in Richmond, California, and is listed on the Nasdaq capital market under the symbol EKSO.

Case Study Questions

1. Discuss long-term strategic investments in technology sectors via public-private partnering
2. Comment on operating in a technology area not in favor with venture capitalists and investment banks
3. What should the role, if any, be of government in robotics technology development?
4. Why are other competing nations making strategic investments in robotics?
5. What was IA's role in connecting the company to private capital markets?
6. Discuss various exit options including a rollup strategy
7. Possibilities of creating a whole new industry
8. Discuss market failures in certain technology areas
9. Should all high-risk technology funding be left to VCs?
10. Handling regulatory issues for startups
11. Take-away(s)

AFTERWORD

The National Science Foundation's Small Business Innovation Program, over the course of more than three decades, has funded thousands of high-technology startups throughout America. In the last ten years or so, each Phase II grantee has been tracked over a period of twelve years from the time of the Phase II award. This has provided insights into various aspects of the culture, vision, management, and risk factors that lead to success, and in many cases failures, of technology startups. A crucial factor, as is expected, is the quality of leadership provided by the founding team. On the federal government side, one important factor involves program policies, flexibilities, and the efficacy of implementation. The program increases its own legitimacy when wise, thoughtful, carefully considered policies implemented are successful and when persons placed in positions of power are seen to have earned their power in a form of meritocracy and demonstrate sufficient wisdom to seem well chosen.

The Table below lists the top ten mistakes that startups usually make:

Developing a product, service, or process that nobody wants
Making mistakes with initial hires
Lack of focus
Failing to properly execute the sales and marketing strategy
Not having the right team of co-founders
Chasing investors and not customers
Not making sure that it has enough money to execute
Spending too much money
Failing to seek help
Not leveraging social media

Change is the engineering process that replaces the fundamental nature of an institution while reform is a medicinal process that preserves the essence while repairing wounds and reviving the essence. Creativity, boldness and effectiveness of decision-making, and a genius for seeking truth from facts, is an important consideration. The distillation of wisdom is a process that normally takes considerable time, and the molding of wisdom into accepted rules by which we can guide our choices takes more time still. Hubris is inherent in human nature – its essence includes prideful overconfidence in the completeness of one's own understanding of the consequences of exercising power in a realm that may well have complexities that extend beyond the understanding of any human.

The values of an organization are criteria by which decisions about priorities are made. Culture is a powerful management tool – culture enables employees to act autonomously and causes them to act consistently. A startup CEO is seen as most effective, if as a leader his/her actions inspire others to dream more, learn more, do more and become more. He/she can articulate with strategic clarity, and to recognize that often it is more important to remove disincentives than to give incentives. Challenges of talent management, role clarity, skill development and attitude management are constant. How to nurture hard skills such as technology development, market analysis, and developing strategic clarity for the business; and soft skills to negotiate relationships, build networks, and find and replace team members.

In a startup, especially, the execution plan is constantly disrupted by new learning. It is driving the train as we are laying the tracks, it is to explore and experiment and then exploit. It is almost a truism that much of the time, groups think better than individuals. Most leaders can expect to have only a few good insights over the course of their careers, and they should not be making moves when they do not have good insights behind them. The willingness to accept the risk of failure is one of the costs of leadership and therefore, the price of all success.

A few closing remarks regarding the evolution and future of the program: Those who suggest something new, should be expected to go to work. Providing a narrow and clear focus for research can be this program's downfall. Prizes targeting specific areas of technology, such as tools and instruments to enable the advancement of fundamental science, can be made part of the NSF SBIR program. It may be useful to consider different solicitation frequencies for different technology segments – for example, solicitations for software companies can be made on a quarterly basis, as opposed to the current six-month cycle regardless of the technology sector. Different timescales for different parts of the same solicitation, and different funding rates too, can be considered. Depending upon the requirements of a particular technology sector, multiple funding cycles for the same startup can be provided.

ACKNOWLEDGEMENT

The author would like to acknowledge the founding team members of the twenty-two startups featured in this volume, for willing to participate in this exercise, of trying to better understand the technology startup landscape. Each provided time and effort to explain over the phone, and in person, during multiple interview sessions, the factors that drive them, and what motivates entrepreneurs, in the first place, to take on a risky endeavor such as creating a technology startup. Numerous thoughtful discussions with NSF SBIR program directors helped hone each narrative. Experts from academia and industry who willingly participate and volunteer their time in proposal review panels at NSF provided additional insights from their unique perspectives.

Finally, I would like to sincerely acknowledge my dear wife, Ligaya S. Nair, for her infinite patience, for helping to edit the final manuscript, and for providing me the time and space, to complete this work.

REFERENCES

1. Descartes, René - Meditations on First Philosophy, 1641

2. Schumpeter, Joseph A. - Capitalism, Socialism and Democracy, 1942

3. Olson, Mancur – The Rise and Decline of Nations: Economic Growth, Stagflation, and Social Rigidities, 1984

4. Burke, Edmund - Thoughts on the Cause of the Present Discontents, 1770

5. Tassey, Gregory – The Technology Imperative, 2009

6. Nair, Murali – New Industry Creation: Discovery to Innovation, 2016

7. Perez, J. C. P. - Active Shape Models with Focus on Overlapping Problem Application to Plant Detection and Soil Pore Analysis, Leibniz University, 2012

8. Fernandez-Perea, M., Larruquert, J. I., Aznarez, J. A., Pons, A., and Mendez, J. A. - Vacuum Ultraviolet Coatings of Al Protected with MgF2 Prepared Both by Ion-beam Sputtering and by Evaporation", Applied Optics, 46, 4871, 2007

9. McDuff, D., R. E. Kaliouby, R. E. et al. - Affect Valence Inference from Facial Action Unit Spectrograms. IEEE Workshop on Analysis and Modelling of Faces and Gestures, 2010

10. Lu, Z. et al. - IEEE Electron Device Letters, Vol. 32, No. 6, pp. 731-733, 2011

11. Oakes F., McTee, S., McMullen, J., Culver, C., Morse, D. - The Effect of Captivity and Diet on KLH Isoform Ratios in Megathura Crenulata, Comparative Biochemistry and Physiology (A) 138, 169-173, 2003

12. Kim, J. - Spray Cooling Heat Transfer: The State of the Art - International Journal of Heat and Fluid Flow, 28(4), 753-767, 2007

13. Global Industrial Control and Factory Automation Market Forecast Until 2016, Published by Markets and Markets, http://www.prweb.com/releases/industrial-control/factory-Automation/prweb9450011.htm., 2013

14. Leonov, V. and R. J. M. Vullers, R. J. M - Wearable Thermo-Electric Generators for Body-powered Devices," Journal of Electronic Materials, 2009

15. Garcia-Romero, A., Diarce, G., Ibarretxe, J., Urresti, A., J.M. Sala - Influence of the Experimental Conditions on the Sub-cooling of Glauber's Salt When Used as a PCM, Solar Energy Materials and Solar Cells, 102, pp 189-195, 2012

16. Simpson, R. - Smart Wheelchairs: A Literature Review, Journal of Rehabilitation Research and Development, Vol. 42, No. 4, pp. 423–438, 2005

17. Alcock, H., White, O., Jegelevicius, G., Roberts, M., and Owen, J. - New High-throughput Methods of Investigating Polymer Electrolytes, Journal of Power Sources, 196 (6), pp. 3355-3359, 2011

18. Report to the Congress on Credit Scoring and Its Effects on the Availability and Affordability of Credit, Board of Governors of the Federal Reserve System, 2007

19. Tarabek, J. et al. - In situ Electron Paramagnetic Resonance (EPR) Spectro-electrochemistry of Single-walled Carbon Nanotubes and C60 Fullerene Peapods, Carbon 44, 2147-2154, 2006

20. Kinniment, D., Heron, K. and Russell, G. - Measuring Deep Metastability – Proceedings of the 12th IEEE International Symposium on Asynchronous Circuits, pp 2-11, IEEE, 2006

21. Bau, O., Poupyrev, I., Israr, A., and Harrison, C. - TeslaTouch: Electro-vibration for Touch Surfaces. In: Proceedings of the 23rd Annual ACM Symposium on User Interface Software and Technology, pp 283-292, New York, New York, 2010

22. Foelix, R. F. - Biology of Spiders; Oxford University Press, New York, New York, 1996

23. Kirsch, N. J., Escobar, E. R., and Turner, R. B. - Antenna Array with Meshed Elements for Beamforming Applications, Proceedings of Progress in Electromagnetic Research Symposium, 2013

24. Lee, J., Chatzichristidi, M., Zakhidov, A., Hwang, H., Schwartz, E., Sha, J., Taylor, P., Fong, H., DeFranco, J., Murotani, E., Wong, W., Malliaras, G. and Ober, C. - Acid-Diffusion Behavior in Organic Thin Films and Its Effect on Patterning. Journal of Material Chemistry, Vol. 19 (19) pp 2986-2992, 2009

25. Dazzi, A., Prazeres, R. Glotin, F., Ortega, J. M. - Subwavelength Infrared Spectro-Microscopy Using an AFM as a Local Absorption Sensor, Infrared Physics and Technology, Volume 49, Issue 1-2, pp 113-121, 2006

26. Meeker J., Sathyanarayana S., and Swan S. - Phthalates and Other Additives in Plastics: Human Exposure and Associated Health Outcomes, Philosophical Transactions of the Royal Society B, 364:2097-2113, 2009

27. Mellado, J., Calmet, J., Olona, M., Esteve, C., Camins, A., Perez, and Del Palomar – Surgically Repaired Massive Rotator Cuff Tears: MRI of Tendon Integrity, Muscle Fatty Degeneration, and Muscle Atrophy Correlated with Intraoperative and Clinical Findings – American Journal of Roentgenology, 184(5), pp 1456-63, 2005

28. H. Kazerooni, H. and Steger, R. - The Berkeley Lower Extremity Exoskeletons, ASME Journal of Dynamics Systems, Measurements and Control, Vol. 128, 2006

ACRONYMS

3D	Three-Dimensional
AAPM&R	American Academy of Physical Medicine and Rehabilitation
ABC	American Broadcasting Company
ADAS	Advanced Driver Assistance System
AFM	Atomic Force Microscopy
AGV	Automated Guide Vehicles
AI	Artificial Intelligence
AMOLED	Active Matric Organic Light-Emitting Diode
ANoC	Autonomic Network-on-Chip
API	Active Pharmaceutical Ingredient
API	Application Programming Interface
ARRA	American Recovery and Reinvestment Act
ASAP	As Soon As Possible
ASIC	Application Specific Integrated Circuit
ASP	Average Sales Price
ATDC	Advance Technology Development Center
ATRS	Automated Transport and Retrieval System
AUTM	Association of University Technology Managers
B2B	Business-to-Business
BEC	Business Email Compromise
BIG	Bayer Innovation GmbH
BPA	Bisphenol-A
CE	Conformite Europeene
CEO	Chief Executive Officer
CES	Consumer Electronics Show
CFO	Chief Financial Officer
CGI	Compacted Graphite Iron
CHiP	Composite High Pressure
CIS	Center for Internet Security
CMO	Contract Manufacturing Organization
CMOS	Complementary Metal Oxide Semiconductor
CMU	Carnegie-Mellon University
CNC	Computer Numerical Control
CNI	Chickasaw Nation Industries
CNSE	College of Nanoscale Science and Engineering
COO	Chief Operating Officer
CPC	Capital Pool Company
CRI	Charlotte Research Institute
CRO	Contract Research Organization
CRRS	Continuous Risk Rating Service
CTO	Chief Technology Officer
DANI	Delay-tolerant Asynchronous Network Interface
DDoS	Distributed Denial of Service
DMA	Dynamic Mechanical Analysis
DNA	Deoxyribonucleic Acid
DOE	Department of Energy
EA	Estrogenic Activity
EBP	Eastman Business Park
ECG	Electrocardiogram
EDA	Electronic Design Automation

EE	Electrical Engineering
EEG	Electroencephalogram
eLEGS	Exoskeleton Lower Extremity Gait System
ELISA	Enzyme-Linked Immunosorbent Assay
EMG	Electromyography
EPA	Environmental Protection Agency
EPR	Electron Paramagnetic Resonance
EQ	Emotional Quotient
ERC	Engineering Research Center
ESR	Electron Spin Resonance
FCC	Federal Communications Commission
FDA	Food and Drug Administration
FICO	Fair Isaac Corporation
FinFET	Fin Field Effect Transistor
FMM	Fine Metal Mask
FPGA	Field Programmable Gate Array
GALS	Globally Asynchronous Locally Synchronous
GE	General Electric
GHz	Gigahertz
GM	General Manager
GM	General Motors
GMP	Good Manufacturing Practices
GNP	Gross National Product
GRC	Global Research Center
GRF	Graduate Research Fellowship
GSR	Galvanic Skin Response
HP	Hewlett-Packard
HPC	High-Performance Computing
HULC	Human Universal Load Carrier
IA	Innovation Accelerator
IC	Integrated Circuit
IEDM	International Electron Devices Meeting
IEN	Institute for Electronics and Nanotechnology
IIP	Industrial Innovation and Partnerships
IoT	Internet of Things
IP	Intellectual Property
IPG	InterPublic Group
IPO	Initial Public Offering
IR	InfraRed
IS	Intrinsically Safe
IT	Information Technology
JDA	Joint Development Agreement
JPL	Jet Propulsion Lab
JVM	Java Virtual Machine
KLH	Keyhole Limpet Hemocyanin
LCD	Liquid Crystal Display
LED	Light-Emitting Diode
M&A	Mergers & Acquisitions
MB	Millward Brown
MBA	Master of Business Administration
MEMS	Microelectromechanical System
MIT	Massachusetts Institute of Technology
MPS	Mail Protection Service

MQL	Minimum Quantity Lubrication
MRI	Magnetic Resonance Imaging
MWF	Metal Working Fluids
NASA	National Aeronautics and Space Administration
NDA	Non-Disclosure Agreement
NFL	National Football League
NHL	National Hockey League
NIH	National Institutes of Health
nm	Nanometer (one-billionth of a meter)
NMS	Natural Microsystems
NNI	National Nanotechnology Institute
NNIN	National Nanotechnology Infrastructure Network
NRE	Non-Recurring Engineering
NSF	National Science Foundation
NYBEST	New York Battery and Energy Storage Technology
NYSERDA	New York State Energy Research and Development Authority
OEM	Original Equipment Manufacturer
OLED	Organic Light-Emitting Diode
ONAMI	Oregon Nanoscience and Microtechnologies Institute
OSC	Organic Semiconductor
OTCBB	Over-The-Counter Bulletin Board
OTFT	Organic Thin-Film Transistor
P&G	Proctor & Gamble
PCT	Patent Cooperation Treaty
PDS	Pallet Detection System
PE	Private Equity
PEEK	Polyether Ether Ketone
PRF	Patient Rehabilitation Facility
PTIR	Photo-Thermal Induced Resonance
R&D	Research and Development
RFID	Radio Frequency Identification
ROI	Return-On-Investment
RTC	Robotics Technologies Consortium
SaaS	Software-as-a-Service
SBIR	Small Business Innovation Research
SCI	Spinal Cord Injury
SDK	Software Development Kit
SDR	Software-Defined Radio
SEC	Securities Exchange Commission
SEMATECH	Semiconductor Manufacturing Technology
SID	Society of Information Displays
SMA	Shape-Memory Alloys
SMP	Shape-Memory Polymers
SNF	Skilled Nursing Facility
SNF	Stanford Nanofabrication Facility
SoC	System-on-Chip
SRAM	Static Random-Access Memory
STEM	Science, Technology, Engineering, and Math
STM	STMicroelectronics
SynBERC	Synthetic Biology Engineering Research Center
TSS	Thermal Storage System
U.S.	United States
UCB	University of California Berkeley

UCSF	University of California San Diego
UI	User-Interface
UM	University of Michigan
UNCC	University of North Carolina, Charlotte
UNH	University of New Hampshire
UV	Ultraviolet
VC	Venture Capital
VMS	Venture Mentoring Service
VP	Value Proposition
VP	Vice-President
WHO	World Health Organization
WISP	Wireless Internet Service Provider
WISPA	Wireless Internet Service Provider Association
WPP	Wire and Plastic Products

www.ingramcontent.com/pod-product-compliance
Lightning Source LLC
Chambersburg PA
CBHW020626220526
45464CB00001B/43